Springerbriefs in Applied Sciences and Technology

Computational Intelligence

W0080218

Series Editor

Janusz Kacprzyk, Systems Research Institute, Polish Academy of Sciences, Warsaw, Poland

SpringerBriefs in Computational Intelligence are a series of slim high-quality publications encompassing the entire spectrum of Computational Intelligence. Featuring compact volumes of 50 to 125 pages (approximately 20,000–45,000 words), Briefs are shorter than a conventional book but longer than a journal article. Thus Briefs serve as timely, concise tools for students, researchers, and professionals.

Pedro H. M. Eid · Filipe P. Azevedo ·
Nuno C. C. Lourenço · Ricardo M. F. Martins

Efficient Analog Integrated Circuit Sizing with GenAI

Exploring Generative Diffusion Models

Springer

Pedro H. M. Eid
Instituto de Telecomunicações
IST-University of Lisbon
Lisbon, Portugal

Filipe P. Azevedo
Instituto de Telecomunicações
IST-University of Lisbon
Lisbon, Portugal

Nuno C. C. Lourenço
Instituto de Telecomunicações
IST-University of Lisbon
Lisbon, Portugal

Ricardo M. F. Martins ⓘ
Instituto de Telecomunicações
IST-University of Lisbon
Lisbon, Portugal

ISSN 2191-530X   ISSN 2191-5318 (electronic)
SpringerBriefs in Applied Sciences and Technology
ISSN 2625-3704   ISSN 2625-3712 (electronic)
SpringerBriefs in Computational Intelligence
ISBN 978-3-031-87104-7   ISBN 978-3-031-87105-4 (eBook)
https://doi.org/10.1007/978-3-031-87105-4

This Springer imprint is published by the registered company Springer Nature Switzerland AG
The registered company address is: Gewerbestrasse 11, 6330 Cham, Switzerland

If disposing of this product, please recycle the paper.

Preface

In recent years the evolution of the electronics industry, and the increase of the demand for consumer-graded products, like smartphones, and computers, has resulted in a reduction in the size and an increase in complexity of IC. While these advancements have propelled the field forward, they have concurrently led the way in a surge of intricacies associated with the design of IC. The challenge of the design can be significant when there is a need for analog parts on the IC, like in a Mixed-Signal System-On-Chip. While the digital aspect has benefited from streamlined design flows through Computer-Aided Design, the field of Electronic Design Automation for analog IC has faced challenges in keeping pace with this complexity. Presently, the conventional design flow for analog circuits requires human intervention due to the absence of automation tools capable of accommodating non-linear components, bias requirements, and real-world effects such as stray impedance. This gap in Computer-Aided Design capabilities has transformed the design of analog circuits into the bottleneck of the IC design. Despite the smaller physical footprint occupied by analog circuits within IC compared to their digital counterparts, the absence of robust automation solutions renders the formers design notably more challenging.

This book contributes to the field of electronic design automation. Its primary focus lies in automating the design of analog integrated circuits, with a particular emphasis on the sizing task of the process. It proposes to leverage ANN, particularly utilizing diffusion models, to enhance and streamline the automation process. Researchers have explored various automation methods, including meta-heuristics and optimization-based approaches, to address this challenge. However, each method presents distinct drawbacks and, at times, yields inefficient results. While studies have made some attempts using ANN, they commonly face the hurdle of the ill-posed nature problem exacerbated by the scarcity of databases for training the models. This work introduces a novel approach based on state-of-the-art generative artificial intelligence to automate the design process, by leveraging diffusion models to enhance the existing ANNs-based framework and address the limitations of previous methodologies. Specifically, Denoising Diffusion Probabilistic Models (DDPM) to tackle the inverse problem of circuit sizing. DDPM employ a noising and denoising architecture, where they learn to reconstruct input distributions by progressively adding and removing noise, starting from Gaussian white noise. Once trained, the DDPM can generate new data from pure noise. Our approach uses this capability to generate new circuit sizing solutions while satisfying performance constraints. The experimental results indicate that our models successfully sized the two tested circuits with an average median error of around 6%. For the smaller circuit in terms of number of sizing design variables, all proposed models surpassed the state-of-the-art approach, whose error was over 60% higher than our models. In the case of the larger circuit, the supervised led to an average error 70% larger than that of our most accurate model. Moreover, by taking advantage of the generative capabilities of our models, we were able to generate points for targets within the dataset, with most of them showing an error below 3%. For the more challenging targets, we managed to find solutions with errors below 10%, while the supervised approaches struggled to achieve errors under 20%.

Finally, the authors would like to express gratitude for the financial support that made this work possible. The work developed in this book was supported by Fundação para a Ciência e a Tecnologia—Ministério da Ciência, Tecnologia e Ensino Superior (FCT/MCTES) through national funds and, when applicable cofounded European Union (EU) funds under the project UIDB/50008/2020 (DOI identifier 10.54499/ UIDB/50008/2020).

This book is organized in six Chapters.

Chapter 1 provides a brief introduction to the problem addressed in this book and its approach.

Chapter 2 reviews the state-of-the-art and various techniques currently being used for analog integrated circuit sizing automation.

Chapter 3 introduces the proposed solution based on generative AI and provides an in-depth description of the different model architectures proposed.

Chapter 4 details the implementation and optimization of the described architectures, including a preliminary evaluation of the models without the use of a simulator.

Chapter 5 discusses the predicted results after evaluating the sizing through a comercial circuit simulator.

Chapter 6 finishes with a conclusion and suggestions for future work on these models.

Lisbon, Portugal   Pedro H. M. Eid
                                                                   Filipe P. Azevedo
                                                                   Nuno C. C. Lourenço
                                                                   Ricardo M. F. Martins

Competing Interests The authors have no competing interests to declare that are relevant to the content of this manuscript.

Contents

Acronyms

ANN	Artificial Neural Network
BO	Bayesian Optimization
CFG	Classifier-Free Guidance
CIPE	Context Independence Estimator
C_{Load}	Capacitance Load
CNN	Convolution Neural Network
cp	Closest point
cv	Closest value
DDPM	Denoising Diffusion Probabilistic Models
ELBO	Evidence Lower Bound
EoT	Encoder-only Transformer
FEI	Finite Expected Improvement
FID	Fréchet Inception Distance
FoM	Figure of Merit
GA	Genetic Algorithms
GANs	Generative Adversarial Networks
GBW	Gain-Bandwidth Product
G_{DC}	DC Gain
GN	Group Normalization
GNN	Graph Neural Network
ICIC	Integrated Circuit
I_{DD}	Bias Current
IS	Inception Score
KL	Kullback-Leibler
ML	Machine Learning
MLP	Multi-Layer Perceptron
MPR	Model Performance Regulator
MSE	Mean Square Error
OPA	Operational Amplifier
PESC	Predictive Entropy Search with Constraints
PM	Phase Margin

PReLU	Parametric Rectified Linear Unit
PSO	Particle Swarm Optimization
PVT	Process-Voltage-Temperature
ResMLP	Residual Multilayer Perceptron
ResNets	Residual Networks
RL	Reinforcement Learning
SL	Supervised Learning
SNN	Shallow Neural Network
VCOTA	Voltage Combiners biased Operational Transconductance Amplifier
VP-SDE	Variance-Preserving Stochastic Differential Equation
wPESC	Weighted Predictive Entropy Search with Constraints

Nomenclatures

μ	Mean of the Distribution
Σ	Variance of the Distribution
σ	Standard Deviation
I	Identity Matrix
T	Total number of time steps
β_t	Controls the distribution of the noise added at each time of the diffusion process
α_t	Represents $1 - \beta_t$
$\bar{\alpha}_t$	Represents the cumulative product of α_t
ϵ_θ	Predicted Noise
ϵ	Gaussian Noise
$q(x_t)$	Perturbed distribution at timestep t
$p_\theta(x_t)$	Predicted distribution at timestep t
$p(x_T)$	Gaussian (Terminal) distribution
$q(x_t|x_{t-1})$	Forward Process Steps
$p_\theta(x_{t-1}|x_t)$	Reverse Process Steps
$\mathcal{N}(x_t; \mu, \Sigma)$	Gaussian distribution describing the behavior x_t
$\mathbb{E}_{x_t,t}$	Expected Value of the joint distribution
∇_{x_t}	Gradient in respect to x_t
$D_{KL}(\cdot|\cdot)$	Kullback-Leibler Divergence

List of Figures

List of Tables

Chapter 1
Introduction

1.1 Motivation

In recent years the evolution of the electronics industry, and the increase of the demand for consumer-graded products, like smartphones, and computers, has resulted in a reduction in the size and an increase in complexity of Integrated Circuit (IC). While these advancements have propelled the field forward, they have concurrently led the way in a surge of intricacies associated with the design of IC.

The challenge of the design can be significant when there is a need for an analog part on the IC, like in Mixed-Signal System-On-Chip and Radio-Frequency IC. While the digital aspect has benefited from streamlined design flows through Computer-Aided Design, the field of Electronic Design Automation for analog IC has faced challenges in keeping pace with this complexity.

Presently, the conventional design flow for analog circuits requires human intervention due to the absence of automation tools capable of accommodating non-linear components, bias requirements, and real-world effects such as stray impedance. This gap in Computer-Aided Design capabilities has transformed the design of analog circuits into the bottleneck of the IC design. Despite the smaller physical footprint occupied by analog circuits within IC compared to their digital counterparts, the absence of robust automation solutions renders the formers design notably more challenging, as exemplified in Fig. 1.1.

Due to this challenging nature, there has been an industry-wide trend to try to replace the analog components of ICs with their digital counterparts. However, the analog elements serve as an essential bridge between the analog of the real world and the digital domain of computer systems. As a consequence, the industry has encountered difficulties in eliminating all the analog ICs due to their role, making their design a challenge but essential for the market.

As of 2024, analog ICs continue to account for over 20% of the total revenue in a market valued at $443.3 billion. And an anticipated growth of $600.1 billion by 2027, reflecting a substantial growth rate of nearly 7%, as delineated in [2]. The size and

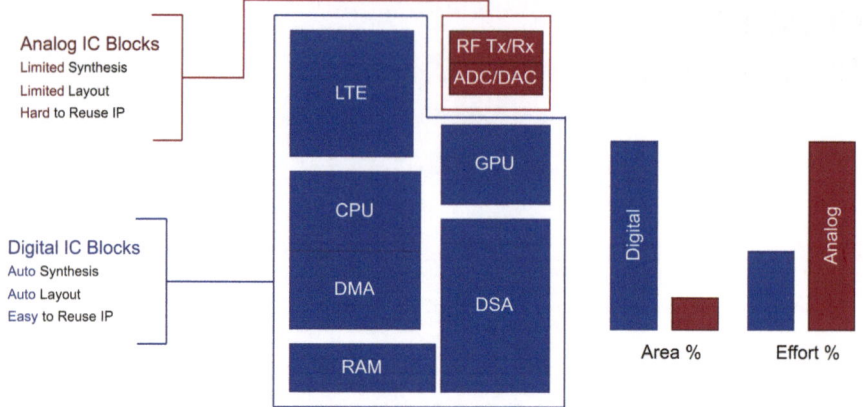

Fig. 1.1 Comparison between design effort and area of the integrated circuits analog and digital parts (adapted from [1])

growth of this market underscores the critical importance of effectively addressing the challenges associated with its design.

1.2   Analog Integrated Circuit Design Flow

The increase of automation in analog IC design has led the way to changes. However, the development of these automation tools still aligns with the sequential process traditionally employed in manual design, that adheres to the steps outlined in [3], illustrated in Fig. 1.2.

The design flow consists of a sequence of top-down steps from the system level to the design level, and bottom-up generation and verification. The top-down methodology facilitates optimization of the system performance, allowing for a more detailed implementation at the device level. Although this approach enhances the likelihood of achieving success in the initial attempts and expedites the design process, real-world factors such as layout parasitic structures or process variations can necessitate multiple iterations to attain a successful design.

The design steps at each level can be divided into distinct processes. The first process is the topology selection where the most appropriate circuit topology that meets a set of given specifications at the current hierarchy level is selected. This selection can either be from existing topologies or synthesized ones. Following this step, there's the Specification Translation, where the specifications of the higher hierarchy block are mapped to the lower levels. This process in lower levels is made by doing circuit sizing, in other words choosing the dimensions (such as widths and lengths) of the smaller components.

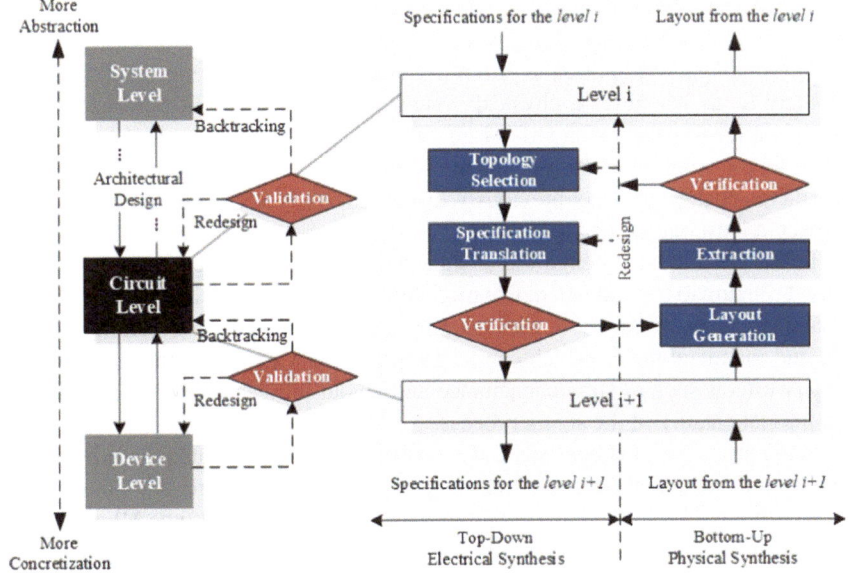

Fig. 1.2 Design flow of analog integrated circuits (reproduced from [3])

The two processes described above are part of the top-down part of the design. The bottom-up phase involves steps such as layout generation. This task involves arranging devices, the dimensions of which were previously determined for the selected topology, within the chip and establishing connections among them. Layout generation consists of two main steps: placement and routing. Placement involves strategically situating devices within the chip area, while the routing phase is the establishment of the interconnections between devices [4].

In the design flow of analog IC, there are still some steps that need to be automated like routing, placement, and circuit sizing. This work will focus on the latter, where given the circuit performance the objective is to determine the sizing of the circuit [5–8].

1.3 Research Goals

In the ever-evolving landscape of IC design, the task of sizing circuits emerges as a critical challenge. Unlike the well-defined process of obtaining the performance from the sizing, where simulations delineate a circuit's performance based on established sizing and topology, circuit sizing involves a reverse process. The goal is to ascertain a circuit topology and sizing values that align with predefined performance metrics.

This inverse problem introduces layers of complexity, often resulting in multiple potential solutions, each with its own set of practical considerations.

Automating the resolution to different types of inverse problems has been a focus in many fields whenever a physical system is to be inferred from measurements. For this issue to have a straightforward solution, it needs to meet three criteria, it is solvable, the solution is unique, and the solution is continuous concerning data and parameter changes [9]. As previously discussed, the design task already does not respect at least one of these criteria.

To simplify this challenge the design is done with a defined topology, leaving only the sizing to be automated. But modeling approaches utilizing traditional mathematics techniques fall short, necessitating a comprehensive consideration of electric and magnetic fields, which can lead to intricate wave equations. With the secondary effects such as stray capacitance and layout intricacies further contribute to the intricate nature of the sizing process.

In the past years, the approaches to resolve this inverse problem have been more focused on the use of Machine Learning (ML) techniques. This book builds upon the existing works in the field [9–11] with a focus on a different type ML technique, namely generative models.

The previously proposed techniques encounter challenges in three main areas. First, approaches that only replace the optimization step are burdened by the high computational and time costs associated with the simulator. Second, offline optimization approaches that bypass the simulator, by reusing available data, can result in suboptimal sizing for various designs that were not adequately represented in the training data. Lastly, the most significant challenge arises from the ill-posed nature. Many ML techniques function essentially as non-linear function approximators, which is why they often struggle when confronted with ill-posed problems with a small dataset [12].

After examining the various approaches discussed in Chap. 2, several shortcomings become apparent. The chosen method for addressing these issues involves using diffusion models, inspired by recent advances in generative models across multiple domains. This approach utilizes diffusion models to tackle the ill-posed inverse problem of circuit sizing.

Generative models provide a robust solution to a range of inverse problems by learning the structure of the dataset, handling noise, and generating probabilistic predictions [13]. Unlike traditional Artificial Neural Network (ANN), these models employ a unique methodology that involves noising and denoising processes to assimilate the dataset's structure and underlying distribution. This novel approach offers significant advantages for the sizing estimation, not being limited to the approximations of this ill-posed problem but learning the underlying distribution of the different solutions.

The goals of this book are outlined below:

- Bypass the manual design or direct optimization process by using a diffusion model that solves the inverse sizing problem offline, eliminating the need for time-consuming online approaches. Once trained, the model can sample new points almost instantly without relying on a simulator.
- Generate new sizing data within the range of the target circuit performance parameters with an improvement when compared to state-of-the-art models of at least 20% error.
- Implement diffusion models for the design of analog ICs for the first time in literature.

1.4 Book Structure

The remainder of this book is organized as follows:

- **Chapter** 2: Reviews the state-of-the-art and various techniques currently in use.
- **Chapter** 3: Introduces the proposed solution and provides an in-depth description of the different architectures proposed.
- **Chapter** 4: Details the implementation and optimization of the described architectures, including a preliminary evaluation of the models without the use of a simulator.
- **Chapter** 5: Discusses the predicted results after evaluating the sizing through a simulator.
- **Chapter** 6: Finishes with a conclusion and suggestions for future work with this model.

References

1. Lourenço, N., Martins, R., Horta, N.: Automatic Analog IC Sizing and Optimization Constrained with PVT Corners and Layout Effects. Springer International Publishing (2017). ISBN: 9783319420370. https://doi.org/10.1007/978-3-319-42037-0
2. Market Insights report. Semiconductors: market data & analysis. In: Statista (2023)
3. Gielen, G.G.E., Rutenbar, R.A.: Computer-aided design of analog and mixed-signal integrated circuits. In: Proc. IEEE **88**(12), 1825–1854 (2000). ISSN: 1558-2256. https://doi.org/10.1109/5.899053
4. Rosa, J.P.S., et al.: Using Artificial Neural Networks for Analog Integrated Circuit Design Automation, pp. 1–8. Springer International Publishing (2020). ISBN: 9783030357436. https://doi.org/10.1007/978-3-030-35743-6
5. Passos, F., et al.: Enhanced systematic design of a voltage controlled oscillator using a two-step optimization methodology. Integration **63**, 351–361 (2018). ISSN: 0167-9260. https://doi.org/10.1016/j.vlsi.2018.02.005

6. Rocha,F.A.E., et al.: Electronic Design Automation of Analog ICS Combining Gradient Models with Multi-objective Evolutionary Algorithms. Springer, Berlin (2014)

7. Mendes, L., et al.: In-depth design space exploration of 26.5-to-29.5- GHz 65-nm CMOS low-noise amplifiers for low-footprint-and-power 5G communications using one-and- two -step design optimization. IEEE Access **9**, 70353–70368 (2021). https://doi.org/10.1109/ACCESS.2021.3078240

8. Póvoa, R., et al.: LC-VCO automatic synthesis using multi-objective evolutionary techniques. In: 2014 IEEE International Symposium on Circuits and Systems (ISCAS), pp. 293–296 (2014). https://doi.org/10.1109/ISCAS.2014.6865123

9. Beaulieu, P.-O., et al.: Analog RF circuit sizing by a cascade of shallow neural networks. IEEE Trans. Comput. Aided Des. Integr. Circuits Syst. **42**(12), 4391–4401 (2023). ISSN: 1937-4151. https://doi.org/10.1109/TCAD.2023.3282570

10. Lourenco, N., et al.: Using polynomial regression and artificial neural networks for reusable analog IC sizing. In: 2019 16th International Conference on Synthesis, Modeling, Analysis and Simulation Methods and Applications to Circuit Design (SMACD). IEEE (2019). https://doi.org/10.1109/SMACD.2019.8795282

11. Lourenco, N., et al.: On the exploration of promising analog IC designs via artificial neural networks. In: 2018 15th International Conference on Synthesis, Modeling, Analysis and Simulation Methods and Applications to Circuit Design (SMACD). IEEE (2018). https://doi.org/10.1109/SMACD.2018.8434896

12. Adler, J., Öktem, O.: Solving ill-posed inverse problems using iterative deep neural networks. Inverse Probl. **33**(12), 124007 (2017). ISSN: 1361-6420. https://doi.org/10.1088/1361-6420/aa9581

13. Mardani, M., et al.: A variational perspective on solving inverse problems with diffusion models (2023). arXiv:2305.04391

Chapter 2
State-of-the-Art

2.1 Exploration of Analog Integrated Circuit Sizing

This section reviews state-of-the-art methods for automating analog Integrated Circuit (IC) sizing, focusing on different methodologies and challenges of the different studies.

2.1.1 Optimization-Based Sizing

Optimization-based sizing can be divided into three main groups: model-based, simulated-based and surrogate-based models. Each of them has different approaches and challenges. The former utilizes approximations of the circuit's performances by trying to find equations, using polynomial equations or macro models (simplified higher-level representations of the behavior of electronic components or circuits). With the model derived the optimization problem can be constructed and solved. The drawbacks with this method are the complexity of the equations that need to be derived and the use of approximations making the prediction less reliable [1]. The simulated-based methods regard the circuit performance as black boxes, using simulators, like *SPICE*, to get the desired outputs. With them, heuristics such as in [2–6], using an evolutionary algorithm, or in [7], using particle swarm optimization, are used to calculate the sizing.

One of the first proposed simulated-based algorithm used in analog IC design challenge is the Genetic Algorithms (GA). Inspired by the principles of natural selection, GA works by iteratively improving a population of candidate solutions. These candidate solutions, analogous to chromosomes in biology, are subjected to selection, crossover, and mutation. The selection process favors solutions with higher

© The Author(s), under exclusive license to Springer Nature Switzerland AG 2025
P. H. M. Eid et al., *Efficient Analog Integrated Circuit Sizing with GenAI*,
SpringerBriefs in Computational Intelligence,
https://doi.org/10.1007/978-3-031-87105-4_2

fitness, as determined by a manually introduced function. That focus on finding better-performing solutions guiding the population towards an optimal design for the sizing problem, as illustrated in Fig. 2.1.

A recent study [8] presents an approach to the GA that focuses on improving the mutation process. This approach aims to address limitations of the traditional GA by guiding the mutation step, leading to a significant reduction in the number of

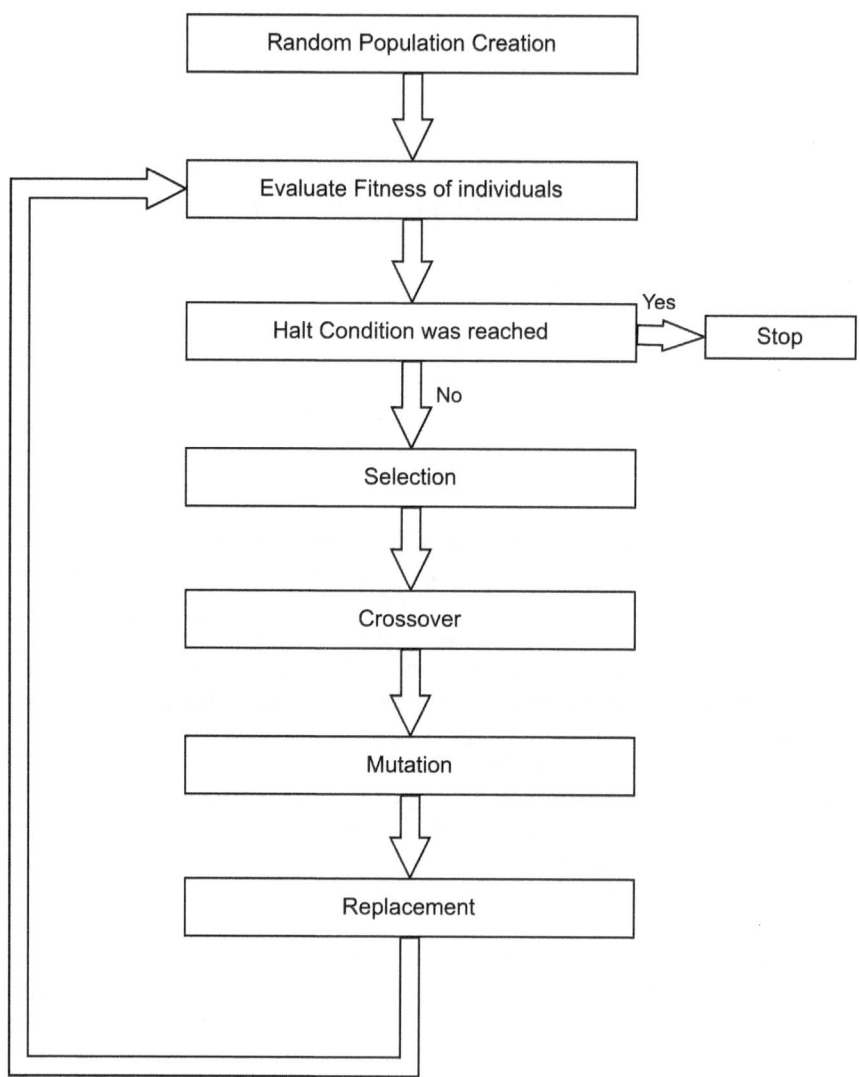

Fig. 2.1 GA process

simulations required to converge on a solution. The authors achieve this by incorporating design rules into the mutation process, as illustrated in Fig. 2.2. This modified
GA demonstrates superior performance when compared to the traditional method.
It achieves a 1.5× and 3.3× faster convergence for a two-stage Operational Amplifier (OPA) and an LC Voltage-Controlled Oscillator, respectively. Furthermore, for
a four-stage OPA design, the standard GA fails to find a suitable solution, while the
modified GA successfully identifies one.

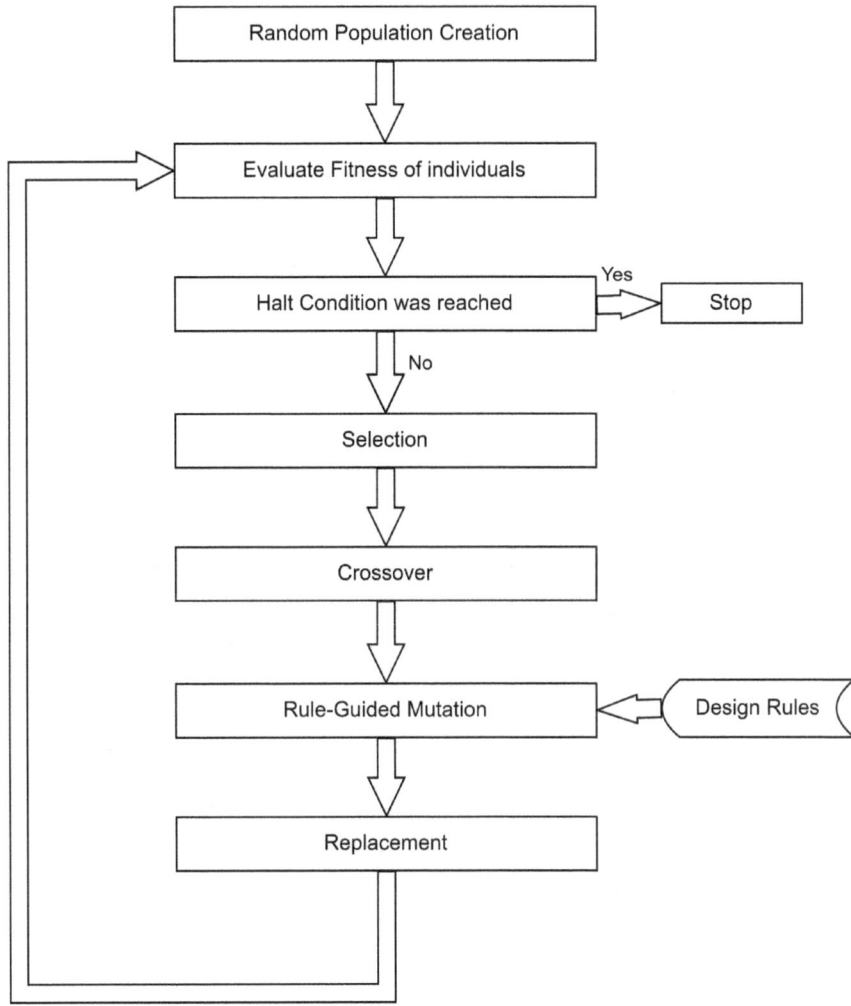

Fig. 2.2 GA process with guided-mutation

Despite these improvements, GA-based approaches remain hindered by substantial computational costs for simulations. Moreover, knowledge gained from one optimization is not directly transferable to subsequent optimizations. Even minor changes in desired solutions or circuit configurations modifications necessitate a completely new optimization with an entirely new set of expensive simulations.

Attempting to bypass the problems of these methods, surrogate methods were proposed. Where it uses both approaches of the previous (model-based and simulated-based) methods to create a new approach. By using an optimization algorithm to add simulated points to iteratively create a model online, using a simulator. This approach can be seen in methods like Weighted Expected Improvement based Bayesian Optimization (BO) in [9], or in a Multi-Task BO [10].

In this part of the section, focus primarily on one technique for the surrogate method, more specifically on BO. The core of Bayesian approaches relies on the Bayes theorem, where prior knowledge is used to create a prior distribution. With this distribution, a posterior distribution is derived by obtaining optimal data with an acquisition function.

One approach using BO can be seen in [1], in this study this surrogate model was used in a multi-test bench approach. Each test bench serves as a simulation environment for testing and verifying the functionality of analog circuit designs. The BO algorithm proposed incorporates two distinct acquisition functions for two different tasks. For the first task, the paper tests two different acquisition functions the Predictive Entropy Search with Constraints (PESC) and its weighted counterpart, weighted Predictive Entropy Search with Constraints (wPESC). They are primarily employed to identify the test bench with the highest level of information. Within a chosen test bench, the exploration ability of PESC/wPESC was seen as insufficient. And to address this limitation, the second acquisition function presented is the Finite Expected Improvement (FEI), which aims to identify the most optimal value within the selected test bench.

This approach yields superior results compared to other surrogate methods mentioned in this section. However, like its counterparts, this approach comes with a significant time cost due to the need for iterative simulations when training its models.

2.1.2   Reinforcement Learning for Integrated Circuit Design

Reinforcement Learning (RL) is a field of Machine Learning (ML) that follows a similar approach as a surrogate model. It does not use label data, instead this algorithm trains online with the simulator by making predictions and receives rewards or punishment depending on the outcome of the prediction. The goal of the training phase is to learn a prediction policy for the actions that maximizes the sum of the rewards.

The use of RL in IC design remains relatively limited, as noted in the literature [11]. A notable recent study in this domain is exemplified in the work presented in [12], where RL is employed to the circuit into distinct sub-blocks, each governed by individual agents.

In this approach, each agent employs two networks: the actor, which handles action decisions, and the critic, which evaluates rewards. Both networks are fully connected with a single hidden layer. The primary distinction between the actor and critic lies in their input and output functions. The actor receives the state, which represents the circuit's performance, and generates the sizing action. Conversely, the critic processes both the action and the state to compute a reward, denoted as Q. Following each action, the actor is trained to adjust his policy in order to minimize Q, thereby improving performance. Furthermore, the paper proposes a refinement to the method by employing two critics instead of one and selecting the one with a lower value. This modification addresses the tendency of the previous approach to overestimate the value function.

Another RL approach for analog IC sizing is detailed in [13]. This study addresses the challenge of the extensive solution space for sizing each component of the IC, which frequently necessitates numerous iterations to find a solution that satisfies performance specifications. To reduce the design space, the paper study utilizing a Graph Neural Network (GNN). This approach begins by converting the circuit net list into a graph, which extracts additional information from the circuit's topology. The graph is then processed using a Relational Graph Convolutional Network (a type of GNN), enabling the network to leverage this topological information to train its heterogeneous weights effectively. Both components of the agent (the critic and the actor) are constructed using this GNN framework.

In addition to utilizing the GNN, the paper puts forth an approach to addressing the challenge of the reward scarcity in the RL field. This approach involves two distinct specifications aimed at enhancing the overall effectiveness of the system. The first specification, set by the designer and related to target performance, is given priority in the optimization process. The second specification involves choosing an arbitrary but reasonable value for parameters such as the area of a capacitor. This serves as a fallback or secondary objective, enabling the system to continue optimization even after the primary specification has been met. Additionally, it is important to note that this RL method employs distinct actions for different sizing parameters.

In a comparative evaluation with optimization methods, BO, the application of RL leads to faster predictions, better accuracy [12], and has better scalability [13]. However, this method also suffers because of the time spent on the simulation.

2.1.3 Supervised Learning for Integrated Circuit Design

The concept of Supervised Learning (SL) in IC design revolves around leveraging datasets containing input and output examples to train a model. After training, this model is employed to forecast new outputs based on new inputs. While diverse techniques exist for implementing SL, emphasis in this section is on the application of Artificial Neural Network (ANN) for this purpose.

The use of SL in different problems has increased dramatically in recent years propelled by an enhancement in computer performance and the availability of vast datasets facilitated the evolution of different methodologies. This surge has significantly enabled the development and utilization of SL across various domains. Nevertheless, when it comes to SL for circuit design, certain limitations persist.

There are two primary challenges in applying SL to circuit sizing. First, as a function approximation method, SL excels at estimating functions, where inputs maps to a unique output. However, circuit sizing is inherently ill-posed, with multiple potential outputs for a given input, necessitating the use of regularization techniques and extensive datasets for accurate function approximation [14]. Second, due to the scarcity of such datasets, SL has primarily been employed to support circuit design, such as replacing simulators or aiding optimizers, rather than directly predicting sizing parameters. Consequently, the field has explored alternative approaches, including augmentation of the dataset and with transfer learning techniques, to address the inverse problem of predicting sizing from performance characteristics.

2.1.3.1   Direct Problem

This section explores the first approach of employing SL for IC sizing. Rather than addressing the sizing estimation directly or devising a new optimization method, this approach centers on reducing simulation or optimization time.

The work presented in [15] focuses on the reduction in simulation time achieved by constructing a dedicated network for each performance parameter and then replacing the simulator with it. These ANNs are trained from data produced in ongoing optimization iterations, receiving the sizing values as input, and the simulation's performance parameters as the validation data. Upon completing the training, the model replaces the simulator, leading to a ∼65% reduction in the execution time. This methodology is illustrated in Fig. 2.3.

In contrast to the aforementioned approach, the work presented in [16] strives to speed up the optimization process. This is achieved by predicting additional input, namely Process-Voltage-Temperature (PVT) corners, for the optimizer through the utilization of sizing from the previous prediction, and the performances for the corresponding tuning mode. In this case, distinct networks are employed to predict PVT corners, with one ANN dedicated to each PVT. This targeted approach serves to constrain the design space for the optimizer, resulting in an accelerated search process for the optimal solution.

2.1.3.2   Inverse Problem

The following section focuses on the exploration of the use of SL to solve the inverse problem. Even with a small dataset, efforts have been made to resolve this challenge. These kinds of approaches focus on replacing the simulator and optimizer for a single

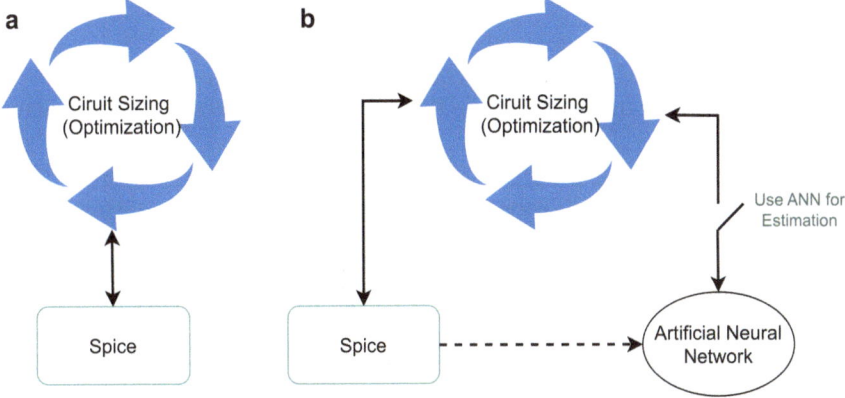

Fig. 2.3 **a** Simulation-based synthesizer, **b** Replacing the simulator with ANN (adapted from [15])

network, directly predicting, from the circuit performance, the sizing of the circuit (e.g., lengths and widths of the MOSFET transistors), as illustrated in Fig. 2.4.

Given the limitations of a small dataset, most of the ANN used to tackle this problem were Shallow Neural Network (SNN). Utilizing a simple Multi-Layer Perceptron (MLP), with few hidden layers, between 3 and 5 [17], or even a single hidden layer [18]. Notably, even using a simple ANN, diverse approaches were explored to effectively tackle this challenge.

On the approach in [17], the primary focus is on the construction of ANN, and the dataset. The network is constructed using a traditional MLP, with Adam optimizer and a few layers, as previously mentioned. The second focus of this paper is on the dataset. The article makes two different approaches to it. The first approach is the use of a dataset from previous optimizations, that contains the only optimal circuit sizing solutions. Second, the dataset is augmented by using the knowledge which circuit performance is often characterized not by a singular value but by inequalities representing a range of values. For example, a solution for a circuit with a performance requirement of S_i is also a valid solution for any circuit with a lower requirement.

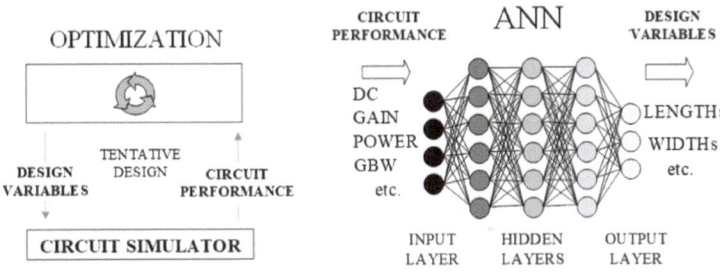

Fig. 2.4 Sizing with ANN (reproduced from [17])

With this information, it was possible to augment the training data with different performance inequalities but with the same sizing objective.

Another approach is seen in [18] that focuses on adding a second element besides the ANN on design automation flow. This proposed automation, which uses two models.

The first part is the Context Independence Estimator (CIPE) which given a target context, uses a Model Performance Regulator (MPR) that optimizes and predicts the circuit performance for the new context. Here, the context refers to conditions like loads, supply voltage, etc. The second part of this chain is the MLP, which predicts the most likely circuit sizing given a target performance, utilizing one hidden layer and the Adam optimizer.

Another recent study done in this field can be seen [19] with a different ANN compared to the other two studies. Instead of relying on a single ANN, it employs a cascade of SNN, as depicted in Fig. 2.5. This architecture utilizes the component sizes of the IC, predicted previously by another network on the cascade as an input for subsequent predictions.

The primary challenge inherent in this architecture lies in the interdependence among components, as the networks rely on previous sizes to predict new components. To resolve the issue, it is also proposed to use a dynamic cascade. The proposed architecture has a similar design to the previous cascade, but when the sizing of one of the components exceeds a certain error threshold, the dynamic mechanism is triggered. The mechanism adds a shallow network at the end of the cascade and then tries again to predict the sizing of the component. The new ANN receives the sizing from the prior networks and their performances, and is permanently integrated into

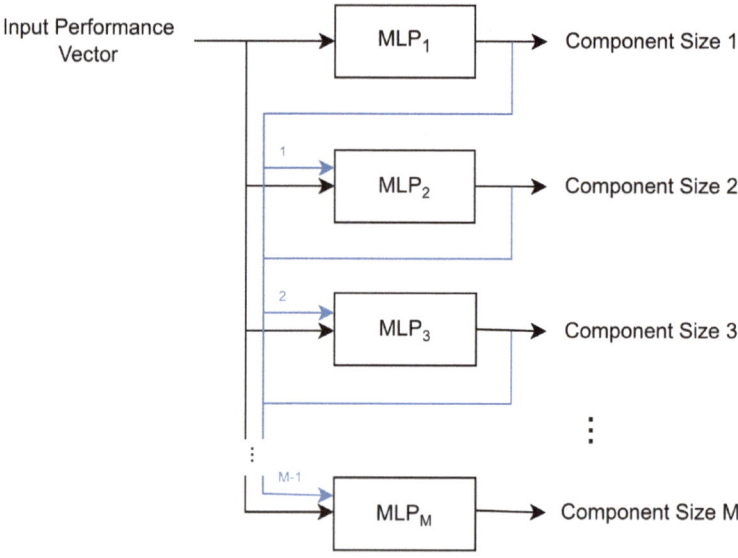

Fig. 2.5 Cascade of SNN (adapted from [19])

the cascade if it can reduce the error of the component. In the updated cascade, more than one network can contribute to the sizing of a component.

2.1.4 Semi-Supervised Learning for Integrated Circuit Design

Semi-supervised learning is a subfield of ML that combines a small amount of labeled data with a large amount of unlabeled data during training. This approach leverages the scarcity of labeled data and the abundance of unlabeled data to improve model accuracy and generalization, addressing the challenges of acquiring extensive labeled datasets.

Although relatively new within the ML paradigm, this subfield offers a valuable alternative to SL for the sizing problem. The high simulation cost of generating all sizing parameters with their respective performance parameters can be mitigated by incorporating unlabeled data into the training process. This approach has been explored in two recent studies.

The first study, [20], adopts an approach similar to [19], utilizing previously predicted sizing to enhance model learning. The design flow is divided into three steps. Initially, different ANNs predict the sizing parameters and, using the performance parameters, derive a validation error for each ANN's output. Based on this error, the predictions are split into two groups: one with smaller errors and one with larger errors. In the second step, an ANN is trained to predict the sizing parameters of the first group. This predicted sizing is then used as input for a second ANN. In the third step, this prediction, along with performance parameters, is used to predict the sizing parameters of the second group. With this approach this study was able to obtain better results than a simple MLP for an OP-AMP with 8 MOSFET@.

Another recent paper focuses on two main areas: utilizing Semi-Supervised Learning to replace the simulator, and subsequently using this replacement to train a hybrid model that combines an evolution-based simulation algorithm (Particle Swarm Optimization (PSO)) with an ANN. Furthermore, the paper simplifies the circuit design process by dividing the circuit into smaller parts, breaking it down into sub-circuits with fewer desired parameters and a reduced solution space. These sub-circuits can then be combined to simulate a more complex circuit using a transfer learning methodology. This approach involves dividing the sub-circuits into body and load components, and then combining different loads and bodies to generate new, more complex circuits. Notably, a small dataset (T_1, T_2, T_3) is used for each of the smaller circuits instead of a dataset for the more complex circuit, resulting in lower computational costs compared to simulating the more complex circuit. The dataset of the complex circuit (T_{out}) is then created from the models of the smaller circuits, with almost zero-computation cost.

This division is managed by implementing a topology decider, which employs three different models. The first model is a classifier trained to generate the specifications of the complex circuit from the specifications of the sub-circuits. The other two models classify the most relevant topologies and generate the low-level specifications from these topologies, respectively.

The sizing prediction process is carried out in two steps. First, an ANN is used for a global search to find the design of the sub-circuit closest to the desired specification. Then, a PSO is used for a local search to predict the sizing. This model was able to maintain the accuracy of predictions with a smaller dataset (from 4.7 to 1090 times smaller) than the state-of-the-art models for one-, two-, and three-stage OP-AMPs, resulting in a faster runtime.

2.1.5   Transfer Learning for Integrated Circuit Design

Transfer learning in ML is a technique widely used across various fields and subfields of ML. This methodology leverages the knowledge gained by a model from one task, particularly the 'unknown knows' of the models, and applies it to a different but related task. Essentially, this technique uses pre-trained models to tackle tasks different from those they were initially trained on. This technique is important when dealing with tasks that have very scarce datasets (Few-shot) or even no data at all (Zero-shot), where obtaining data is difficult. Transfer learning becomes highly valuable in scenarios where a similar task with abundant data is available. In IC sizing, transfer learning can be leveraged in two primary ways. As circuit become more complex, their simulations also become more computationally expensive, whereas simulations of simpler circuits are faster and more data can be more readily obtained. The first implementations focus on using models trained on simpler and different topologies to predict parameters for circuits of another topology. This approach was employed and described in the paper [21] from the previous section, although it was not the paper's primary focus.

A study that primarily addresses this methodology is [22]. In this study, a supervised model was developed to predict performance from sizing using a GNN. Four different circuit topologies were used: a two-stage op-amp, a three-stage op-amp, a folded cascode op-amp, and a telescopic cascode op-amp. The model was trained on three of these topologies and then applied to predict the sizing of the fourth topology using transfer learning techniques.

Two transfer learning techniques were evaluated. The first, a Few-shot approach, involved freezing most layers of the model and retraining only the post-processing and final layers of the GNN with a small dataset (1%) from the fourth circuit. The second, a Zero-shot approach, utilized the model trained on the three other circuits to make predictions for the fourth circuit without any additional data. Both approaches were tested against multiple MLPs, each trained on a different circuit with 1% of the respective dataset. The Zero-shot approach generally outperformed the basic ANN in most performance metrics, while the Few-shot approach achieved superior results across all metrics.

A simpler and more conventional transfer learning approach in IC sizing involves applying knowledge across different technologies rather than different topologies. The paper [23] explores this by using a simple ANN to transfer knowledge from a two-stage operational amplifier in a 180 nm process to a 65 nm process. This study employs a Few-shot approach and examines the impact of fine-tuning versus freezing different layers during retraining.

In this method, the model is initially trained with all available data from the 180 nm process. It is then retrained with less than 10% of the data from the 65 nm process. For comparison, a standalone model was also retrained with 10% of the 65 nm data. The transfer learning approach yielded better results than the standalone model, with a lower test loss (less than 10%). Various configurations of frozen layers were tested, revealing a trade-off between accuracy and retraining time: fewer frozen layers reduced error but increased retraining time.

A recent implementation that focus on the use of transfer learning techniques to predict the sizing can be found in the paper [24]. This study focuses in utilizing the dataset of smaller components of the circuit like i.e., current mirror, different pair, to train smaller models and transfer this knowledge to a more complex circuits sizing models. The main approach of the paper is to create a network-database of more simple circuit ANN. Instead of using a new big dataset to train a new circuit network, it used these pre-trained models and small set of this new circuit dataset to fine-tune the model.

The networks developed in this paper are capable of predicting the ratio between width and length for different OP-AMPs based on this network-database. This approach was able to produce robust predictions for Folded Cascode OP-AMP and a Telescopic topology, achieving less than 5% error for most of the performance parameters.

2.1.6 Analog Integrated Circuit Sizing Automation Summary

See Table 2.1.

2.2 Exploration of Diffusion Models

In this part of the chapter, diverse approaches and applications of diffusion models will be explored, a category of generative ANN that are primarily utilized in the realms of image generation and restoration. The rest of this first section focuses on a generic implementation of the diffusion model.

They were first formulated in [26], by applying concepts of thermodynamics, specifically drawing inspiration from the non-equilibrium thermodynamics. This

Table 2.1 Overview of the state of the art of analog integrated circuit sizing

Reference	Method	Results
Jafari et al. [25]	Traditional GA algorithm	It was able to find solution to different two-stage OPA
Zhou et al. [8]	Modified GA with guided mutation	The method surpassed GA in speed and successfully designed four-stage OPAs
Zhao et al. [1]	Two-step BO method with wPESC and FEI, one to select the bench and another to find the solution inside the bench	When compared to state-of-art optimization-based methods it has a higher sampling efficiency and thus, speed-up
Zhang et al. [12]	Multi-agent Reinforcement-Learning with different MLP for each actor and critic	Compared to Bayesian Optimization methods results in faster, higher accuracy
Li and Carusone [13]	Reinforcement-Learning using Graph representation for the IC with GNN for the actor and critic	Compared to single-agent RL using MLP, it is faster and more accurate, and it scales better than BO
Islamoglu et al. [15]	Simulator Replacement with SPEA2 algorithm (simulation-based)	Achieving a $\sim$ 65% reduction in the execution time and an average of error .4% when compared to the simulator
Vaz et al. [16]	Tries to speed the optimization by predicting the PVT, with various MLP, bypassing the simulator	Obtained a 2.97$\times$ speed when compared to the same problem without ANN
Lourenco et al. [17]	MLP with 3–5 layers, and optimal or quasi-optimal augmented dataset	It was able effectively to do sizing of an IC within the range of parameters in almost an hour
Lourenco et al. [18]	Two-step method utilizing multivariate polynomial regression to estimate the trade-off between the new context, with the output of it being used as an input of MLP	Taking less than 15 s to train, it was able to size with an error of $\sim$1%
Beaulieu et al. [19]	Fixed and Dynamic cascade of MLPs	Took DC-SNN 4 min and mixer (up to 1 h) in training. Both predict in < 15 sec with MAE < 0.05 for Dynamic cascade
Mina et al. [20]	A two-stage ANN system was designed to predict sizing parameters, dividing the task into simpler and more complex sub-problems	For an OP-AMP with 8 MOSFET it obtained a 3.5 times smaller validation loss than a single ANN
Fayazi et al. [21]	Uses Semi-Supervised Learning and Transfer Learning to train a simulator of sub-circuits to then train hybrid system of PSO and an ANN	Tested in 1st,2nd and 3rd stage OP-AMP with average MAPE of 0.027 for all circuit and smaller dataset than the state of the art
Wu and Savidis [22]	Few-Shot and Zero-Shot approaches for the direct problem	Compared to MLP, few-shot was the best overall, while zero-shot often performed better
Wu and Savidis [23]	Compared few-shot and standalone approaches on two-stage op-amp transfer from 180 to 65 nm, testing different layer freezing configurations.	Transfer learning improved accuracy over standalone training, with accuracy increasing alongside training time as more layers were fine-tuned
Leibl and Graeb [24]	Creates a network-database of the simple circuits, and use this database to fine-tune for different OP-AMPs	Sizing a Folded Cascode and Telescopic topology OP-AMP with most performance targets with a smaller error than 5%

methodology consists primarily of three steps. The first two steps, denoted as the forward and reverse, are leveraged to train the model. And the last process, the sampling process is utilized to generate the new data.

The forward process is characterized by T sequential steps where in each step t the data x_t is perturbed by adding a small noise ϵ_t. These steps are described by the distribution $q(x_t|x_{t-1})$. Following the completion of all steps, what remains of the original data $q(x_0)$ manifests as the terminal distribution $q(x_T)$. This process has no trainable parameter, and it can be expressed by the Eq. 2.1.

$$q(x_T|x_0) := q(x_1|x_0)\ldots q(x_T|x_{T-1})$$
$$= \prod_{t=1}^{T} q\,(x_t \mid x_{t-1}) \tag{2.1}$$

The reverse process entails training a denoising ANN to iteratively eliminate the noise from x_T in order to recover the initial data. By utilizing the terminal distribution, this process does the steps in the reversal order until it reaches the initial step. Described by the distribution $p_\theta(x_{t-1}|x_t)$, where θ represents the trainable parameters of the network, and the entire process can be encapsulated by the Eq. 2.2.

$$p_\theta(x_0) := p(x_T)p_\theta(x_{T-1}|x_T)\ldots p_\theta(x_0|x_1)$$
$$= p(x_T)\prod_{t=1}^{T} p_\theta(x_{t-1}|x_t) \tag{2.2}$$

These networks are usually trained using the Evidence Lower Bound (ELBO) on the negative log-likelihood, with the Kullback-Leibler (KL) method, by computing the statistical distance, in other words measuring how the predicted probability $(p_\theta(x_t))$ distribution diverges from the real distribution $(q(x_t))$. After the training, the final process, the sampling, begins by acquiring new data from the terminal distribution (a pure noise). In sequence, the process leverages the optimal ANN to denoise the pure noise. Resulting in the generation of new data from the initial distribution. In a nutshell, this is accomplished by effectively rerunning the reverse process with the now-optimized parameters of the ANN.

With a generic understanding of Diffusion Models, the rest of this part of the chapter focuses on understanding the different approaches of this method, ranging from the formulation of models to the different approaches for each process.

2.2.1 Formulations

There are two main formulations of diffusion models. The first is the discrete formulation, where the steps of the diffusion models are defined by discrete values from 0 to T. In contrast, the continuous formulation has the steps defined as continuous values inside a boundary.

2.2.1.1   Discrete Formulation

The most common model for the discrete diffusion model, and the focus of this section, is the Denoising Diffusion Probabilistic Models (DDPM) [27], this model is easy to train and gives high-quality results. DDPM employs a forward process that involves adding Gaussian noise, as depicted in Eq. 2.3.

$$
\begin{aligned}
q(x_t|x_{t-1}) &:= \mathcal{N}(x_t; \mu, \Sigma) \\
&:= \mathcal{N}(x_t; \sqrt{1 - \beta_t}x_{t-1}, \beta_t I)
\end{aligned}
\tag{2.3}
$$

With β as the hyperparameter of the forward process, governing the amount of noise that is added. Since this process just adds Gaussian noise, all the different steps can be obtained from the initial with the Eq. 2.4, with $\overline{\alpha_t}$ being the product of all the previous $1 - \beta_t$.

$$
q(x_t|x_0) := \mathcal{N}(x_t; \sqrt{\overline{\alpha_t}}x_0, (1 - \overline{\alpha_t})I)
\tag{2.4}
$$

The second process of DDPM, the reverse, removes the Gaussian noise, and differently than the forward process, can't have all its steps parameterized by a single equation. The different steps are represented by Eq. 2.5.

$$
\begin{aligned}
p_\theta(x_{t-1}|x_t) &= \mathcal{N}(x_{t-1}; \mu_\theta(x_t, t), \Sigma_\theta(x_t, t)) \\
&:= \mathcal{N}\left(x_{t-1}; \frac{1}{\sqrt{\alpha_t}}\left(x_t - \frac{1 - \alpha_t}{\sqrt{1 - \overline{\alpha_t}}}\epsilon_\theta(x_t, t)\right), \beta_t I\right)
\end{aligned}
\tag{2.5}
$$

Here, α_t represents $1 - \beta_t$. The network of the DDPM is trained by finding the closest learned noise (ϵ_θ) to the real noise (ϵ_t), illustrated by the Eq. 2.6.

$$
\mathcal{L} = \mathbb{E}_{x_t, t}\left[\left\|\epsilon_t - \epsilon_\theta\left(\sqrt{\overline{\alpha_t}}x_0 + \sqrt{1 - \overline{\alpha_t}}\epsilon, t\right)\right\|_2^2\right]
\tag{2.6}
$$

2.2.1.2   Continuous Formulation

The continuous formulation manipulates the dataset in continuous time, whereby formulating the steps of the processes of diffusion model as infinitesimal intervals, and creates stochastic differential equations, to represent them.

A notable example of this formulation model is the Variance-Preserving Stochastic Differential Equation (VP-SDE) from [28], which will be the focus of this section. In this example, both the forward process and the reversed process are defined by the Eq. 2.7, respectively.

$$dx = -\frac{1}{2}\beta(t)x dt + \sqrt{\beta(t)}dw$$
$$dx = -\frac{1}{2}\beta(t)x dt - \beta(t)\Delta_{x_t} \log(p(x_t)) + \sqrt{\beta(t)}d\bar{w} \qquad (2.7)$$

The w and $\bar{w}$ denote the Wiener and reverse Wiener processes, respectively, serving as the stochastic components in the differential equations. While the construction of the reverse and forward processes shares similarities, the reverse equation introduces an additional term related to the score function, representing the parameters that are trained in this process.

An alternative perspective on these models is explored in [28] with the model RED-Diff. In this study, the challenges of image generation and in-painting are framed as a reverse problem. Specifically, the task involves determining the initial data x_0 using a measurement model f with knowledge of the observations y, which can be nonlinear and subject to noise v.

The resolution of this issue is accomplished through sampling from a pre-trained variational diffusion model. This relationship is illustrated by Eq. 2.8.

$$y = f(x_0) + v \qquad (2.8)$$

The proposed variational model uses KL minimization with a new variational distribution. The optimization aims to reduce the statistical distance between two distributions. The article simplifies the optimization into the Eq. 2.9.

$$\min_{\mu} \left\{ \underbrace{\|y - f(\mu)\|^2}_{\text{recon}} + \lambda_t \underbrace{(\text{sg}\,[\epsilon_\theta(x_t;t) - \epsilon])^T \mu}_{\text{reg}} \right\} \qquad (2.9)$$

Addressing the optimization above entails determining an image, denoted as μ, which effectively reconstructs the observed data under the measurement model. Simultaneously, this reconstruction strives to maintain a high likelihood under the constraints set by the prior, as enforced by the regularization term. The regularization term's impact is modulated by the weight parameter λ_T, a hyperparameter responsible for striking a balance between the influence of the prior and the likelihood.

In comparison to analogous models such as ΠGDM, $DDRM$, and DPS, the methodology employed in RED-Diff [28] proves superior in generating more refined images.

2.2.2 Forward Process

The primary objective of the forward process is to perturb the initial dataset in order to attain the terminal distribution. Various approaches have been proposed to achieve this goal, and this section delves into the diverse schedules and types of noise employed in this initial process.

A suitable schedule encourages a balance between exploration and exploitation. Exploration describes the ability to generalize data not seen during training while exploitation refers to the convergence situation where a model fits the training data well. A sufficient amount of noise is necessary to encourage exploration to generalize well on unseen data, while excessive noise may result in a model that cannot adequately recover the details of data. On the other hand, too little noise boosts exploitation to fit distributions well but undermines generalization [29].

While there have been attempts to leverage ANNs for determining the most optimal schedule, manually designed noise schedules featuring mathematical equations have found broader applications. Among them, the use of sigmoid, cosines, and mathematical integrals has superior results when compared to exponential and linear, which tend to add noise too quickly, making it difficult to find optimal solutions. The comparison is illustrated in Fig. 2.6.

The choice of noise type further contributes to enhance the distribution approximations. Gaussian Noise, in its various forms, is commonly utilized for its simplicity. Additionally, the use of a Gamma distribution represents a feasible and promising type of noise for fitting the distribution.

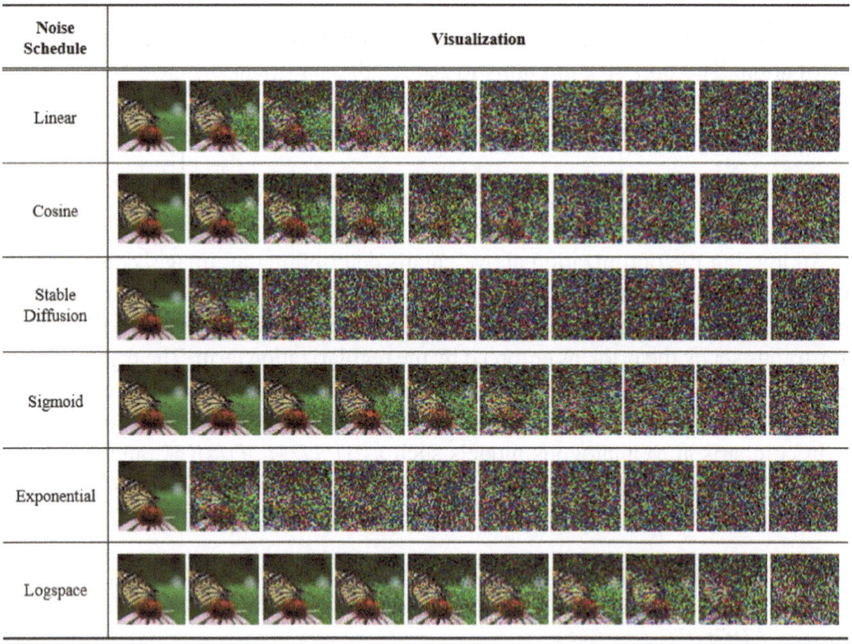

Fig. 2.6 Noise schedule (adapted from [29])

2.2.3 Reverse Process

The reverse process aims to train an ANN to reconstruct the initial dataset from the terminal distribution. Various approaches have been proposed to achieve this objective, with this section specifically examining the architecture of the ANN and the parameterization of the Reverse Mean.

The two widely employed denoising networks are the U-Net and Transformers. The U-Net can be characterized as a U-shaped network of encoder-decoders, fostering effective feature learning. On the other hand, Transformers constitute a network of encoder-decoders with self-attention, promoting scalability and multi-modality. Both of these architectures have gained widespread adoption due to their ability to model complex relationships within datasets.

When addressing the steps of the reverse process (2.5), learning the reverse mean (μ_θ) becomes crucial. Three primary approaches are commonly used for this object. The first involves computing the mean from approximations of the initial data, a straightforward yet potentially inaccurate method in the later stages of the process. Another approach is to approximate μ_θ–by introducing a score function ($s_\theta(x) = \nabla_x \log(p_\theta(x))$), calculated as the gradient of the logarithm of the distribution of x at time step t, representing the most probable changes between two time steps. The final approach to learn the mean is by approximating the noise (ϵ_x) used in the forward process. While challenging at the process's outset, this approach proves advantageous as it progresses toward the end.

2.2.4 Sampling Process

The final stage of the diffusion model, the sampling process, encompasses various aspects explored through diverse studies. This section, however, will primarily focus on the different approaches to the guidance mechanism. The guidance mechanisms play a pivotal role in denoising the sampling process and establishing conditions for the model, particularly in problems like text-to-image. After training on a dataset, the guidance mechanism becomes instrumental in steering the denoising of the sampling process.

One simple approach to guidance is the use of vanilla guidance, illustrated by the Eq. 2.10, which incorporates the conditions in each time step t of the process.

$$p_\theta(x_T|y) = p(x_T) \prod_{t=1}^{T} p_\theta(x_{t-1}|x_t, y) \tag{2.10}$$

With the y representing the encoded text from a text-to-image model, or low-resolution image to perform super-resolution [30]. However, it lacks adjustable conditional weight and the studies [31] show that it may not always conform to the conditions.

An alternative strategy involves the utilization of a classifier to guide the diffusion model. The fundamental concept revolves around employing a classifier trained on the diffusion model's training data. By sampling the gradient at each classification, it becomes feasible to integrate this term into the score estimation sampling, resulting in the formulation of Eq. 2.11:

$$\tilde{\epsilon}_\theta(x|y) = \epsilon_\theta(x) + w\nabla_{x_t} \log p_\theta(y \mid x_t, t) \qquad (2.11)$$

While the classifier-guided approach provides enhanced control through adjustable weights, it introduces additional costs associated with training the classifier. Moreover, given the inherent noise in the training dataset, this method may result in instability.

A more prevalent alternative is the utilization of Classifier-Free Guidance (CFG). This approach aims to achieve the guidance without relying on the gradient of the classifier. This is achieved by training both unconditional ($p(x)$) and conditional models ($p(x|y)$), akin to the vanilla guidance. The relation between both reverse mean estimates can be defined as $\epsilon_\theta(x) = \epsilon_\theta(x|y = \varnothing)$. Using this definition, both models are jointly trained by introducing a new hyperparameter, p_{uncond}, which randomly sets the class identifier to $\varnothing$, thus, training the unconditional model [32]. Ultimately, this sampling process involves superposing both models to create a new one where the guidance of the diffusion model is controlled by an adjustable weight (w). This eliminates the necessity for a trained classifier and mitigates potential instability associated with classifier-guided methods. Additionally, jointly training the models minimizes the added time and cost. The resulting model is mathematically represented in Eq. 2.12.

$$\tilde{\epsilon}_\theta(x|y) = (1 + w)\epsilon_\theta(x|y) - w\epsilon_\theta(x) \qquad (2.12)$$

2.2.5 Evaluation Metrics

The improvements of generative ANN for synthesis of images has had an increase in the public interest. Evaluating these models using traditional ML metrics, such as validation loss, presents challenges because these models need to maintain fidelity to the training set while also generating diverse images without merely memorizing them.

The use of metrics such as Fréchet inception distance (FID) and Inception Score (IS) has become standard for evaluating generative models, including Generative Adversarial Networks (GANs) and Diffusion models. Both metrics rely on a pretrained Convolution Neural Network (CNN), specifically the Inception-V3 network. This CNN maps images into a latent space and in this latent space a score is calculated. For instance, the FID computes the Fréchet distance between the latent spaces of generated and real images. These metrics allow models to be assessed in terms of both the diversity and fidelity of the generated images, offering a more comprehensive evaluation compared to traditional metrics.

2.2.6 Stable Diffusion

To finish this exploration of diffusion models, one of the most influential implementations of diffusion models, Stable Diffusion, will be discussed. This model can be seen as the one that brought significant attention to these types of generative models, thanks to its effective implementation and open-source availability. It was first introduced in the paper [33]. The model builds on previous configurations of diffusion models, particularly the discrete configurations.

Building on the implementations of DDPM, the paper introduces some different implementations. There are three significant modifications. First, following the implementation of diffusion denoising implicit models, it introduces a non-Markovian chain. With this chain, it is able to sample images in smaller time steps than the ones used during training, while still retaining the information from the trained time steps. This is implemented to reduce the time taken to sample from the model without compromising the quality of the generated images. Second, instead of directly working in the pixel space, it uses an Autoencoder to encode the pixels into a lower-dimensional space, which improves the scalability of the model. Finally, it makes changes to the U-Net used for denoising the images. Transformer encoders are added to tokenize the text, which is then used to guide the denoising process. With these tokens, a cross-attention mechanism is employed to incorporate this information into the time-steps.

2.3 Conclusion

In this chapter, a comprehensive investigation was conducted into the automation of the analog IC sizing process and diffusion models. Diverse methodologies for automating sizing were examined, each characterized by distinct approaches and associated challenges. By juxtaposing these explorations, this study helps to illustrate an effective implementation of the diffusion model to generate sizing datasets for IC. The challenges of the state-of-art of circuit design can be summarized in these points:

- Both RL and BO typically require online training, which involves numerous interactions with the environment. In the context of circuit sizing, these 'interactions' are expensive simulations that need to be run multiple times. The high computational cost and time-consuming nature of these makes tuning and training these models expensive.

- SL methods, and related techniques, perform best when there is a well-defined mapping between input features and unique output targets. However, in the context of analog circuit sizing, this is an ill-posed problem because the relationship between design parameters (outputs) and performance metrics (inputs) is neither straightforward nor unique. This complexity often leads to issues such as overfitting and poor generalization, as the model may struggle to accurately capture the underlying mapping in the data. Although these challenges can be mitigated by an extensive datasets and regularization techniques [34], the scarcity of datasets in this domain exacerbates these issues, making the limitations of such approaches even more apparent.

The choice of the model is related to these two challenges, but it focuses primarily on the second one. Diffusion models offer an alternative by shifting the focus from learning a direct input-output mapping to learning the distribution of possible outputs given an input. Specifically, diffusion models are capable of understanding the distribution of possible outcomes and using guidance parameters to direct this distribution. This probabilistic approach aligns well with the inherent ill-posed nature of analog circuit design, making diffusion models a good alternative for this problem. Moreover, since they are trained offline, they eliminate the need for expensive simulations each time the models requires tuning.

References

1. Zhao, J., et al.: A novel and efficient bayesian optimization approach for analog designs with multi-testbench. In: 2022 27th Asia and South Pacific Design Automation Conference (ASP-DAC). IEEE (2022). https://doi.org/10.1109/ASP-DAC52403.2022.9712590
2. Liu, B., et al.: Analog circuit optimization system based on hybrid evolutionary algorithms. Integration **42**(2), 137–148 (2009). ISSN: 0167-9260. https://doi.org/10.1016/j.vlsi.2008.04.003
3. Passos, F., et al.: Enhanced systematic design of a voltage controlled oscillator using a two-step optimization methodology. Integration **63**, 351–361 (2018). ISSN: 0167-9260. https://doi.org/10.1016/j.vlsi.2018.02.005
4. Rocha, F.A.E., et al.: Electronic Design Automation of Analog ICS Combining Gradient Models with Multi-objective Evolutionary Algorithms. Springer, Berlin (2014)
5. Mendes, L., et al.: In-depth design space exploration of 26.5-to-29.5- GHz 65 nm CMOS low-noise amplifiers for low-footprint-and-power 5G communications using one-and- two -step design optimization. IEEE Access **9**, 70353–70368 (2021). https://doi.org/10.1109/ACCESS.2021.3078240
6. Póvoa, R., et al.: LC-VCO automatic synthesis using multi-objective evolutionary techniques. In: 2014 IEEE International Symposium on Circuits and Systems (ISCAS), pp. 293–296 (2014). https://doi.org/10.1109/ISCAS.2014.6865123
7. Vural, R.A., Yildirim, T.: Analog circuit sizing via swarm intelligence. AEU-Int. J. Electron. Commun. **66**(9), 732–740 (2012). ISSN: 1434-8411. https://doi.org/10.1016/j.aeue.2012.01.003
8. Zhou, R., Poechmueller, P., Wang, Y.: An analog circuit design and optimization system with rule-guided genetic algorithm. IEEE Trans. Comput.-Aided Des. Integr. Circuits Syst. **41**(12), 5182–5192 (2022). ISSN: 1937-4151. https://doi.org/10.1109/TCAD.2022.3166637

9. Lyu, W., et al.: An efficient Bayesian optimization approach for automated optimization of analog circuits. IEEE Trans. Circuits Syst. I: Regul. Pap. **65**(6), 1954–1967 (2018). ISSN: 1558-0806. https://doi.org/10.1109/TCSI.2017.2768826

10. Huang, J., et al.: Bayesian optimization approach for analog circuit design using multi-task gaussian process. In: 2021 IEEE International Symposium on Circuits and Systems (ISCAS). IEEE (2021). https://doi.org/10.1109/ISCAS51556.2021.9401205

11. Mina, R., Jabbour, C., Sakr, G.E.: A review of machine learning techniques in analog integrated circuit design automation. Electronics **11**(3), 435 (2022). ISSN: 2079-9292. https://doi.org/10.3390/electronics11030435

12. Zhang, J., et al.: Automated design of complex analog circuits with multiagent based reinforcement learning. In: 2023 60th ACM/IEEE Design Automation Conference (DAC). IEEE, San Francisco, CA, USA (2023)

13. Li, Z., Carusone, A.C.: Design and optimization of low-dropout voltage regulator using relational graph neural net work and reinforcement learning in open-source SKY130 process. In: 2023 IEEE/ACM International Conference on Computer Aided Design (ICCAD). IEEE (2023). https://doi.org/10.1109/ICCAD57390.2023.10323720

14. Kamyab, S., et al.: Survey of deep learning methods for inverse problems. (2021). arXiv: 2111.04731 [cs.CV]

15. Islamoglu, G., et al.: Artificial neural network assisted analog IC sizing tool. In: 2019 16th International Conference on Synthesis, Modeling, Analysis and Simulation Methods and Applications to Circuit Design (SMACD). IEEE (2019). https://doi.org/10.1109/SMACD.2019.8795293

16. Vaz, P., et al.: Speeding-up complex RF IC sizing optimizations with a process, voltage and temperature corner performance estimator based on ANNs. In: 2022 IEEE International Symposium on Circuits and Systems (ISCAS). IEEE (2022). https://doi.org/10.1109/ISCAS48785.2022.9937911

17. Lourenco, N., et al.: On the exploration of promising analog IC designs via artificial neural networks. In: 2018 15th International Conference on Synthesis, Modeling, Analysis and Simulation Methods and Applications to Circuit Design (SMACD). IEEE (2018). https://doi.org/10.1109/SMACD.2018.8434896

18. Lourenco, N., et al.: Using polynomial regression and artificial neural networks for reusable analog IC sizing. In: 2019 16th International Conference on Synthesis, Modeling, Analysis and Simulation Methods and Applications to Circuit Design (SMACD). IEEE (2019). https://doi.org/10.1109/SMACD.2019.8795282

19. Beaulieu, P.-O., et al.: Analog RF circuit sizing by a cascade of shallow neural networks. IEEE Trans. Comput. Aided Des. Integr. Circuits Syst. **42**(12), 4391–4401 (2023). ISSN: 1937-4151. https://doi.org/10.1109/TCAD.2023.3282570

20. Mina, R., Sakr, G.E., Nassif, H.: Enhancing transistor sizing in analog IC design using a circuit-focused semi-supervised learning. In: 2023 IEEE 4th International Multidisciplinary Conference on Engineering Technology (IMCET). IEEE (2023). https://doi.org/10.1109/IMCET59736.2023.10368264

21. Fayazi, M., et al.: AnGeL: fully-automated analog circuit generator using a neural network assisted semi-supervised learning approach. IEEE Trans. Circuits Syst. I: Regul. Pap. **70**(11), 4516–4529 (2023). ISSN: 1558-0806. https://doi.org/10.1109/TCSI.2023.3295737

22. Wu, Z., Savidis, I.: Transfer learning for reuse of analog circuit sizing models across technology nodes. In: 2022 IEEE International Symposium on Circuits and Systems (ISCAS). IEEE (2022). https://doi.org/10.1109/ISCAS48785.2022.9937457

23. Wu, Z., Savidis, I.: Transfer learning for reuse of analog circuit sizing models across technology nodes. In: 2022 IEEE International Symposium on Circuits and Systems (ISCAS). IEEE (2022). https://doi.org/10.1109/ISCAS48785.2022.9937457

24. Leibl, M., Graeb, H.: Optimizer-free sizing of OpAmps leveraging structural and functional properties. In: 2024 20th International Conference on Synthesis, Modeling, Analysis and Simulation Methods and Applications to Circuit Design (SMACD). IEEE (2024)

25. Jafari, A., et al.: Design of analog integrated circuits by using genetic algorithm. In: 2010 Second International Conference on Computer Engineering and Applications. IEEE (2010). https://doi.org/10.1109/ICCEA.2010.118
26. Sohl-Dickstein, J., et al.: Deep unsupervised learning using nonequilibrium thermodynamics (2015). arXiv:1503.03585
27. Ho, J., Jain, A., Abbeel, P.: Denoising diffusion probabilistic models (2020). arXiv:2006.11239
28. Mardani, M., et al.: A variational perspective on solving inverse problems with diffusion models (2023). arXiv:2305.04391
29. Chang, Z., Koulieris, G.A., Shum, H.P.H.: On the design fundamentals of diffusion models: a survey (2023). arXiv:2306.04542
30. Understanding diffusion models: a unified perspective (2022). https://www.calvinyluo.com/2022/08/26/diffusion-tutorial.html#guidance
31. Luo, C.: Understanding diffusion models: a unified perspective (2022). arXiv:2208.11970
32. Ho, J., Salimans, T.: Classifier-free diffusion guidance (2022). arXiv:2207.12598
33. Rombach, R., et al.: High-resolution image synthesis with latent diffusion models (2021). arXiv:2112.10752
34. Adler, J., Öktem, O.: Solving ill-posed inverse problems using iterative deep neural networks. Inverse Probl. **33**(12), 124007 (2017). ISSN: 1361-6420. https://doi.org/10.1088/1361-6420/aa9581

Chapter 3
Proposed Architectures

3.1 Denoising Diffusion Probabilistic Model Overview

In this section, an overview of the Denoising Diffusion Probabilistic Models (DDPM) is presented, as initially presented in [1] and subsequently improved in [2]. The DDPM utilizes a discrete diffusion formulation with Gaussian noise, establishing the foundational framework for the approaches discussed in this book.

As discussed in the preceding chapter, the DDPM involves three distinct processes. The forward and reverse processes are represented by Eqs. 2.4 and 2.2, respectively. Furthermore, the individual steps of the denoising process are shown in Eq. 2.5. The last process, sampling, is described by the same equations as the reverse process but employs an optimized Artificial Neural Network (ANN). A visual representation of the model is presented in Fig. 3.1.

This model was chosen because it is straightforward to define, efficient to train, and capable of achieving high quality and high diversity in the generated samples [2]. The implementations of both the training and sampling are illustrated in Fig. 3.2

In this model, the estimation of the reverse mean (μ_θ) involves an approximation with noise (ϵ_θ). The improved DDPM [2] introduces two pivotal modifications. Firstly, it integrates cosine scheduling instead of a linear schedule. Secondly, it parameterizes the reverse variance according to Eq. 3.1, demonstrating significant improvements in sampling quality.

$$\Sigma_\theta(x_t, t) = \exp(v \cdot \log(\beta_t) + (1 + v) \cdot \log(\tilde{\beta}_t)) \tag{3.1}$$

Here, β_t and $\tilde{\beta}_t$ are hyperparameters, where $\tilde{\beta}_t = \frac{1-\bar{\alpha}_{t-1}}{1-\bar{\alpha}_t}\beta_t$ represents its scaled-up counterpart. This parametrization represents a learned linear interpolation between these values. The modification introduces a variation in Eq. 2.6, incorporating a term for the newly learned parameter v, which is related to the lower bound variance described in the Eq. 3.2. Here, L_{simple} is described by the Eq. 2.6, and the mean in the variance part is considered a constant, without the calculation of the gradient of

P. H. M. Eid et al., *Efficient Analog Integrated Circuit Sizing with GenAI*,
SpringerBriefs in Computational Intelligence,
https://doi.org/10.1007/978-3-031-87105-4_3

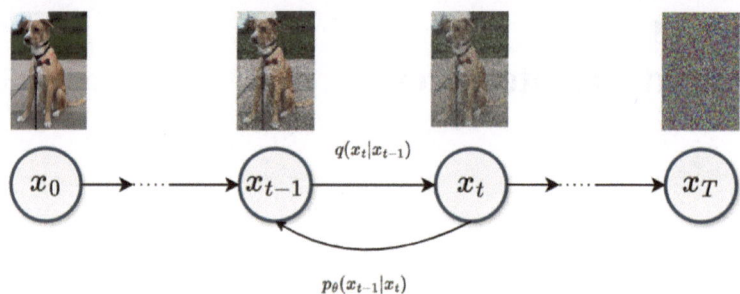

Fig. 3.1 Denoising diffusion probabilistic model schematics

Algorithm 1 Training Loop

1: **repeat**
2:    $x_0 \sim q(x_0)$
3:    $t \sim \mathcal{U}(1, \ldots, T)$
4:    $\epsilon \sim \mathcal{N}(0, I)$
5:    **Compute Gradient:**
6:    $\nabla_\theta \left\| \epsilon - \epsilon_\theta \left(\sqrt{\overline{\alpha}_t} x_0 + \sqrt{1 - \overline{\alpha}_t} \epsilon, t \right) \right\|^2$
7: **until** converged

Algorithm 2 Sampling Process

1: $x_T \sim \mathcal{N}(0, I)$
2: **for** $t = T \ldots, 1$ **do**
3:    $z \sim \mathcal{N}(0, I)$ **if** t>1 else z=0
4:    $x_{t-1} = \frac{1}{\sqrt{\alpha_t}} \left(x_t - \frac{1-\alpha_t}{\sqrt{1-\overline{\alpha}_t}} \epsilon_\theta(x_t, t) \right) + \sigma_t z$
5: **end for**
6: **return** x_0

Fig. 3.2 DDPM pseudocode

the term. Additionally, λ is a hyperparameter, typically set to $\lambda = 0.001$, to avoid overwhelming the mean part of the loss.

$$
\begin{aligned}
L_{vlb} &= L_0 + L_1 \cdots + L_T \\
L_{t-1} &= D_{KL} \left(q(x_{t-1}|x_t, x_0) \| p_\theta(x_{t-1}|x_t) \right) \\
L_{hybrid} &= L_{simple} + \lambda L_{vlb}
\end{aligned}
\tag{3.2}
$$

In the reverse process, a commonly utilized ANN architecture is the U-Net. Widely recognized for its effectiveness in image analysis, the U-Net features two cascaded Convolution Neural Networks (CNNs), establishing an encoder-decoder architecture. The encoder facilitates the down-sampling of input images and extracts high-level features, while the decoder performs up-sampling to reconstruct the output. These components are interconnected with skip-connections that facilitate the flow of information across different segments of the architecture.

Also, for the architecture proposed in the paper [2] there are two additional layers that enhance stability and performance. The first is an Adaptive Group Normalization (GN) within the CNN. This normalization technique builds upon GN by dynamically selecting groups of data from image channels for standardization. Secondly, a Multi-Head Attention layer is inserted between the CNNs and GN, in the decoder part of the network. This attention layer helps the model focus on the most relevant features and better learn the different time steps of the diffusion model.

3.1.1 Guidance Overview

The following section provides an overview of the Classifier-Free Guidance (CFG) approach and its implementation. This form of guidance harnesses the diffusion model to simultaneously learn both conditional and non-conditional distributions, employing them to guide the final distribution.

Originally presented in [3], this approach addresses the challenges seen in the classifier guidance introduced in [2]. The Classifier guidance complicates the diffusion model training pipeline because it requires training an extra classifier, and this classifier must be trained on noisy data, so it is generally not possible to plug in a pre-trained classifier [3].

A distinguishing feature of the CFG approach is its use of the DDPM itself as a substitute for the classifier. The training process of the DDPM with CFG is illustrated in pseudocode in Fig. 3.3.

By sampling from the conditional and unconditional score estimates, defined in Eq. 2.12, the CFG method actively directs the sampling process of DDPM. The specific adjustments to this final process are detailed in the pseudocode presented in Fig. 3.4.

Implemented within DDPM, CFG plays a fundamental role in guiding the model, leveraging text embeddings to influence image generation. This approach introduces two new hyperparameters, p_{uncod} and w, where in [4] the values of 0.1 and 0.3 were chosen, respectively.

Algorithm 3 Training Loop with Classifier-Free Guidance

1: **repeat**
2:   $x_0 \sim q(x_0)$
3:   $c = y$ if $\mathcal{U}(0,1) > p_{uncod}$ else $\emptyset$
4:   $t \sim \mathcal{U}(1,\dots,T)$
5:   $\epsilon \sim \mathcal{N}(0,I)$
6:   **Compute Gradient:**
7:   $\nabla_\theta \left\| \epsilon - \epsilon_\theta \left(\sqrt{\overline{\alpha_t}} x_0 + \sqrt{1 - \overline{\alpha_t}} \epsilon, t, c \right) \right\|^2$
8: **until** converged

Fig. 3.3 Training pseudocode with classifier-free guidance

Algorithm 4 Sampling Process with Classifier-Free Guidance

1: $x_T \sim \mathcal{N}(0, I)$
2: **for** $t = T \ldots, 1$ **do**
3:     $\epsilon_\theta(x_t, t, c) = (1 + w)\epsilon_\theta(x_t, t, y) - w\epsilon_\theta(x_t, t, \emptyset)$
4:     $z \sim \mathcal{N}(0, I)$ **if** t>1 **else** z=0

5:     $x_{t-1} = \frac{1}{\sqrt{\alpha_t}}\left(x_t - \frac{1-\alpha_t}{\sqrt{1-\bar{\alpha}_t}}\epsilon_\theta(x_t, t, c)\right) + \sigma_t z$

6: **end for**
7: **return** x_0

Fig. 3.4 Sampling pseudocode with classifier-free guidance

3.2 Denoising Diffusion Probabilistic Model for Sizing

This section will delve into one distinct theme, it will explore the adaptation of the DDPM to incorporate vector representations of sizing and performance, particularly the ANN employed in this model. The approach involves employing a CFG to guide the denoising of the sizing vector, utilizing the vector of performance values. This will serve as the primary strategy for tackling the problem. A simplified diagram of this model can be observed in Fig. 3.5.

For this approach, the proposed DDPM incorporates a cosine schedule to address the sizing problem. It also integrates noise estimation for the reverse mean and introduces a parameterized reverse variance, similar to the first implemented version in the initial paper [1]. By directly parameterizing with β_t, the complexity of the ANN decreases. While parametrization following the (3.1) equation could improve the results, it might also lead to a time-inefficient model, as the model would need to predict both the noise estimate and v.

The fundamental distinction between the sizing problem and the challenges commonly encountered with diffusion models lies in the structure of the datasets. While

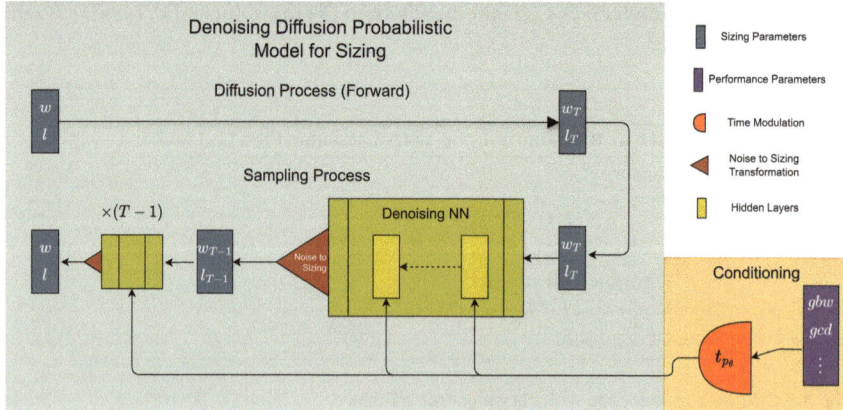

Fig. 3.5 Diagram of the DDPM for sizing

the diffusion models typically receive matrices (images) as inputs for the DDPM, the sizing problem presents a unique scenario with vectors as the datasets. Consequently, certain adjustments are necessary to effectively incorporate this type of information. The modification is made in the type of ANN in the reverse process. This book will focus primary on constructing of the DDPM with CFG and different types ANN.

3.2.1 Proposed Guidance

As discussed earlier, the CFG plays crucial part on the diffusion model by guiding the generation, and it is the main approach of the solution for sizing. This approach involves a transformation step that ensures compatibility with the network's layer sizes and performance vector. This vector of performances information is then modulated in the time step embedding (detailed later on this chapter). The resulting combined signal is subsequently added to the different layers of the network, as illustrated in the schematic Fig, 3.6.

This guidance step is influenced by the hyperparameter p_{uncond}. This parameter determines whether the model is trained conditionally (with performance information) or unconditionally (without the performance). In simpler terms, p_{uncond} determines whether the performance gets incorporated into the time step embedding. During the sampling stage, after the training, the other hyperparameter w plays a crucial role in balancing the influence of the conditional and unconditional models in final prediction.

3.2.2 Artificial Neural Network

This section focuses on discussing the different ANN used in conjunction with the DDPM to address the inverse problem of sizing analog Integrated Circuits (ICs).

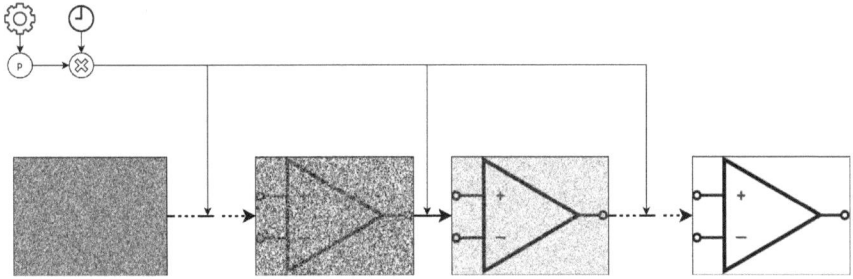

Fig. 3.6 Classifier-free guidance schematic (adapted from [5])

3.2.2.1   Fully-Connected Neural Network

The first proposed ANN adopts an Multi-Layer Perceptron (MLP), as depicted in
Fig. 3.7, with a dynamic number of layers and neurons determined during optimization. This MLP serves as the ANN for the reverse process, replacing the U-NET.

The fully connected network is structured to process a vector that contains the
sizing features with an added noise of the different time steps. This vector serves as the
input, and the network's goal is to learn efficient noise prediction across the different
time steps. The specific configuration of neurons and layers is determined through
the optimization. Each hidden layer within this architecture follows a consistent
structure, as depicted in Fig. 3.8. The specific architectural choices for each block in
these hidden layers will be further explored in this chapter.

3.2.2.2   Fully-Connected Neural Network with Skip Connections

Another proposed architecture for addressing this problem integrates concepts from
both Residual Networks (ResNets) and U-NETs, introducing skip connections into
the MLP. This approach uses a residual networks that incorporates the information
from earlier layers into subsequent layers within the network. By doing so, it facilitates the recovery of information that might be lost across the network depths and
alleviates the issue of the vanishing gradients.

This architecture, termed Residual Multilayer Perceptron (ResMLP), maintains
a similar structure to the previous MLP, featuring a variable number of layers and
neurons, and adheres to the hidden layer structure shown in Fig. 3.8. However, in
each layer beyond the middle one, the output of earlier layers is added to the input of
subsequent layers. This method aims to enhance the learning process and improve
the performance. A schematic of this network is depicted in the Fig. 3.9.

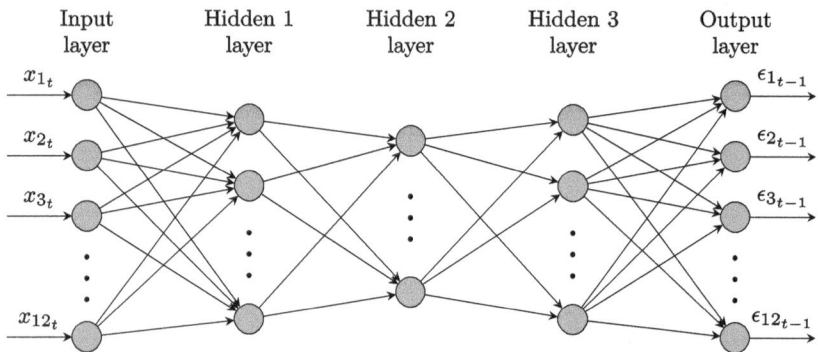

Fig. 3.7 MLP example

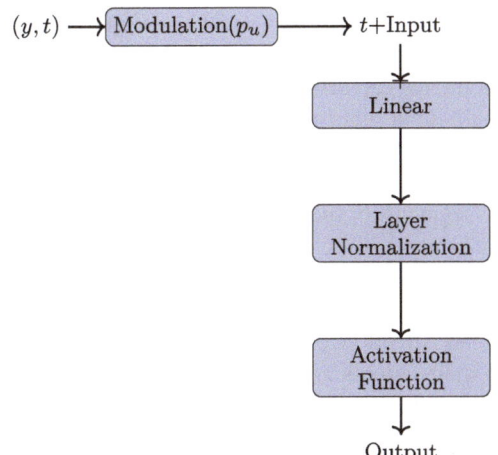

Fig. 3.8 Feed-forward hidden layers

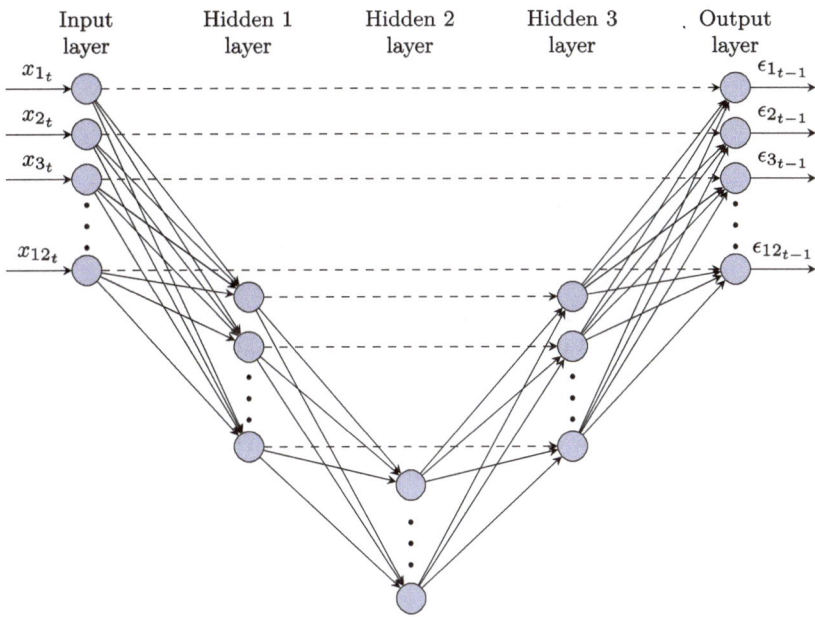

Fig. 3.9 MLP with skip connections

3.2.2.3 Encoder-Only Transformer

The last implemented architecture for this problem draws inspiration from the transformer architecture, particularly the encoder part.

The key component in the transformer, and what differentiates this proposed architecture from other ANNs in this chapter, is the Multi-Head Self-Attention mechanism. This mechanism facilitates capturing dependencies and relationships between different positions within input sequences. It has become a foundational component in various neural network architectures, initially popularized by the transformer model paper [6]. This mechanism enables processing sequences in parallel and selectively attending to diverse segments of the input.

The fundamental concept of Self-Attention involves computing and optimizing three different matrices: query, key, and value. These matrices are used to calculate attention scores that help the model focus on the most relevant information. In the case of the Multi-Head Attention, each head performs the self-attention mechanism independently, and their outputs are then concatenated. This approach allows the model to further focus on the most important parts of the input by calculating attention score multiple times.

Self-Attention, specifically Multi-Head Attention, is also prominently utilized in the U-NETs of the DDPM, as previously mentioned, where it plays a crucial role in enhancing the model's ability to capture dependencies and conditional information across different time steps.

Taking as reference the transformer and encoder-only models like BERT [7], a modified Encoder-only Transformer (EoT) block was implemented for this problem.

This network maintains the same architecture between the blocks as the previous ResMLP Fig. 3.9, including skip connections. On the other hand the Feed-Forward layer follows the design of the previous proposed hidden layers Fig. 3.8, incorporating a linear layer, layer normalization, and an activation function. It is worth mentioning that the Multi-Headed Attention module includes an output layer that retains the input dimensionality, which is crucial for this block's residual connections.

Two main modifications distinguish this architecture from classical implementations. First, the second residual connection was excluded due to unsatisfactory results. This could be explained by the fact that the second connection forces the network to maintain the dimensionality of the sizing input, compressing the perceptron output throughout the entire network. This constraint might have led to the loss of important information, given the small dimensionality of the input. The second major modification was the removal of the first layer normalization block, which also led to unsatisfactory results.

3.2.3 Common Architectures Choices for the Neural Networks

Each of the different proposed ANN architectures share common structures that facilitate their integration and training with the DDPM. In this section, the focus will be on five main architectural choices: the Time Embedding, Conditional Modulation, Layer Normalization, and Activation Function and Optimizer.

3.2.3.1 Time Embedding

The Sinusoidal Positional Embedding was chosen as the time embedding mechanism. This approach was first introduced for the Transformer architectures in the paper [6]. The sinusoidal embeddings are defined using sine and cosine functions of different frequencies. The equations for the positional encoding are as follows:

$$
\begin{aligned}
PE_{(pos,2i)} &= \sin\left(\frac{pos}{10000^{\frac{2i}{d}}}\right) \\
PE_{(pos,2i+1)} &= \cos\left(\frac{pos}{10000^{\frac{2i}{d}}}\right)
\end{aligned}
\tag{3.3}
$$

where pos is the position, i is the dimension index, and d is the dimensionality of the embeddings. These equations ensure that the time encoding for each time-step is unique and that similar time-steps have similar embeddings, which helps the model understand the order and relative distances between different positions in the time-step. After embedding it, the time information is added to the different layers of the networks.

3.2.3.2 Conditional Modulation

The main approach for the sizing problem is to use the performance parameters to modulate the time embedding, guiding the denoising process of the proposed model. Various methods were tested to incorporate this conditional information into the signal.

The first approach follows the typical method used in DDPM, where both the time embedding and performance parameters are passed through separate feed-forward networks to produce vectors of the same dimension as the hidden layer. Then are added together, which the result is passed to the input of the hidden layer. The second approach is similar but involves an element-wise multiplication of the time and performance vectors, instead of addition. And the final approach, inspired by the method tested in [8], combines the previous two methods. It involves an affine transformation of the time embedding by producing two different vectors, each with a different perceptron, from the performance parameters: one for the element-wise multiplication and the other for the addition, this approach is described in the Eq. 3.4. The implementation and evaluation of these different approaches are explored in the subsequent chapter.

$$
t_y = \gamma_y \odot t_{emb} + \beta_y
\tag{3.4}
$$

3.2.3.3 Layer Normalization

Layer normalization was a critical implementation for the model. While the Feed Forward Networks without it were able to produce decent results, the networks tended to become unstable as the number of layers increased. To address this issue, layer normalization across the different hidden layers was introduced. This technique standardizes the output of the different layers, thereby maintaining consistent mean and variance, and promoting a more stable and efficient training. The formula below describes this transformation.

$$\text{LayerNorm}(x_i) = \frac{x_i - \mu}{\sqrt{\sigma^2 + \epsilon}} \cdot \gamma + \beta \qquad (3.5)$$

where the μ and σ are the expected values and standard deviation respectively. And γ and β are learnable parameters of the network, that allows the network to scale and shift the distribution to better fit the output.

It is worth noting that Batch Normalization was also considered during testing for the feed-forward networks. However, given that Layer Normalization follows a similar approach to GN, which is frequently utilized in the U-NET of DDPM, and also does not depend on the batch size, this type of normalization was selected.

3.2.3.4   Optimizer and Activation Functions

It is important to note that the choice of activation function and optimizer remained consistent across all models. Initially, ReLU was utilized for the activations. However, as the models were optimized, it was discovered that achieving satisfactory results required a deeper network architecture. Consequently, Parametric Rectified Linear Unit (PReLU) [9], an extension of the ReLU function designed to mitigate issues such as 'dying ReLUs', was selected. This phenomenon occurs when neurons become inactive during training due to the fixed zero slope for negative values, a common challenge in deep neural networks that was observed in the initial implementations. In the PReLU, an additional learnable parameter introduces a small slope for negative inputs, effectively addressing the problem caused by ReLU's fixed zero slope. The distinction between ReLU and PReLU is illustrated in Fig. 3.10.

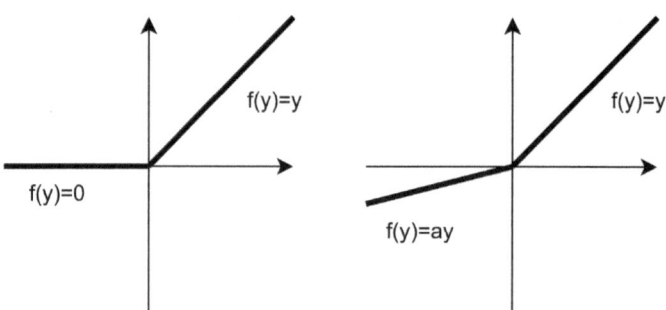

Fig. 3.10 Comparison between ReLU and PReLU

The optimization process employed in this study utilized the Adam optimizer. This optimizer is commonly chosen due to its adaptive learning rate capability, which adjusts based on the gradients' first and second moments. This adaptation offers a more efficient convergence than classical optimizers like stochastic gradient descent.

3.3   Evaluation

This section focuses on discussing the evaluation and comparison methodologies used to assess the proposed models.

3.3.1   Auxiliary Neural Network

Training a diffusion model poses significant challenges, particularly in assessing its performance. The conventional metrics, such as validation and test losses, often fall short in reflecting the true prediction errors due to the multiple denoising steps involved. Also, the generative nature of the model and the ill-defined nature of the inverse problem make direct comparisons of sizing predictions challenging. Consequently, adopting an alternative evaluation strategy becomes essential.

In the literature, diffusion models are commonly evaluated using metrics such as Inception Score (IS) or Fréchet inception distance (FID), which gauge the diversity of generated images by leveraging the Inceptionv3, a pre-trained CNN, to establish a latent space for images. However, the proposed model does not generate images, making it unsuitable for direct evaluation using such methods.

Additionally, constructing an ANN to derive a latent space for circuit sizing outputs and employing similar metrics would be ineffective. The goal is to size circuits for specific performance targets, a task complicated by the ill-posed nature that can contain multiple solutions.

To streamline model optimization across the different hyperparameters without relying on time-consuming simulations, an auxiliary network is proposed. This network is tailored to address the task of obtaining performance metrics from generated circuit sizing and with that compare with target performance, facilitating the evaluation of the different models configurations. Importantly, this auxiliary network is employed solely for comparative purposes. After optimizing the model, the final predictions are directly evaluated using a simulator.

The auxiliary network is constructed using a simple MLP, similar to the first described ANN of the DDPM, with a dynamic number of layers and neurons. It takes the sizing as input (s) and predicts the performance (p) of the circuit, observed on the Fig. 3.11, providing a more efficient way to evaluate the diffusion model.

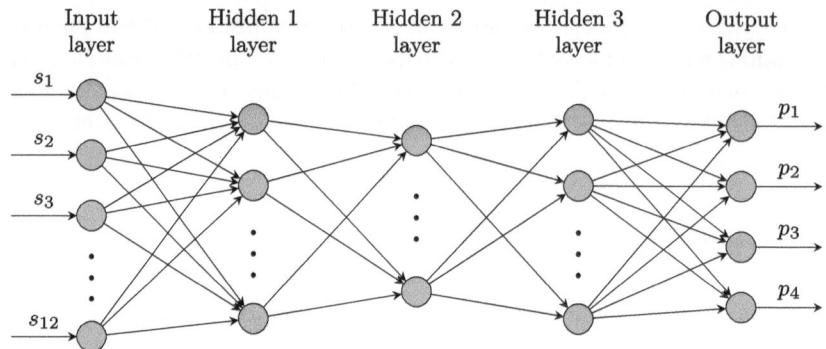

Fig. 3.11 Simulator replacement for evaluation

3.3.2 Supervised Leaning Comparison

To provide a benchmark for the different approaches, a traditional MLP was constructed and trained using supervised learning methodology. This MLP adopted a structure similar to the auxiliary ANN of Fig. 3.11 but instead of predicting performance based on sizing parameters, it aimed to determine sizing from given performance metrics.

3.4 Optimization

To optimize the different ANN a Bayesian Optimization (BO) was employed using the Optuna Python library, detailed in [10]. Optuna automates hyperparameter tuning by testing various configurations and utilizes algorithms like the Tree-structured Parzen Estimator (a type of BO algorithm) to efficiently explore the search space. It incorporates pruning techniques to halt unpromising trials early, thus conserving computational resources.

This tool facilitated optimization across the various ANN architectures and played a pivotal role in optimizing the w hyperparameter in the guidance mechanism of DDPM.

3.5 Conclusion

This chapter proposed various architectures for using a DDPM to address the ill-posed inverse problem of circuit sizing. First, the base architecture of the DDPM along with its different configurations were discussed. This was followed by an examination of the architecture of the denoising ANN. Finally, a discussion on the evaluation, along with a benchmark model for comparison, and optimization metrics and tool was provided. The subsequent chapter will discuss the use of this optimization metric and tool to find the most optimal configuration of the proposed models.

References

1. Ho, J., Jain, A., Abbeel, P.: Denoising diffusion probabilistic models (2020). arXiv:2006.11239
2. Nichol, A., Dhariwal, P.: Improved denoising diffusion probabilistic models (2021). arXiv:2102.09672
3. Ho, J., Salimans, T.: Classifier-free diffusion guidance (2022). arXiv:2207.12598
4. Mongaras, G.: Diffusion models—DDPMs, DDIMs, and classifier free guidance. Medium (2023)
5. Chang, Z., Koulieris, G.A., Shum, H.P.H.: On the design fundamentals of diffusion models: a survey (2023). arXiv:2306.04542
6. Vaswani, A., et al.: Attention is all you need (2017)
7. Devlin, J., et al.: BERT: pre-training of deep bidirectional transformers for language understanding (2018). arXiv:1810.04805
8. Perez, E., et al.: FiLM: visual reasoning with a general conditioning layer (2017). arXiv:1709.07871
9. He, K., et al.: Delving deep into rectifiers: surpassing human-level performance on imagenet classification (2015). arXiv:1502.01852
10. Akiba, T., et al. Optuna: a next-generation hyperparameter optimization framework (2019). arXiv:1907.10902

Chapter 4
Model Implementation and Optimization

4.1 Dataset

This section describes the datasets used to train the models and the feature engineering tools applied to these datasets.

4.1.1 Description

The proposed method using the diffusion model will be tested on different analog circuit topologies, such as the Voltage Combiners biased Operational Transconductance Amplifier (VCOTA) and a Folded VCOTA, these circuits are illustrated in Fig. 4.1a and b, respectively. The different sizing of the transistors (represented in Fig. 4.1a and b) of these Integrated Circuit (IC)s will be used as the input. They will help the model to learn the structure of each IC dataset through the noising and denoising process, enabling the creation of new realistic sizing from a white Gaussian noise. It is important to note that these datasets weren't collected by me.

The VCOTA has 12 different transistors, organized into 6 pairs of matched transistors. The dataset for this circuit contains pairs of widths and lengths for each pair. For example, transistors 0 (MP0) and 1 (MP1) has their sizing defined by W_0 and L_0. The Folded VCOTA, on the other hand, contains 8 matched transistors, with each pair having the width, length per finger, and the number of fingers. It also includes 3 additional transistors, each with 3 additional sizing parameters, related to V_{BN} and V_{BP}. In summary, the VCOTA contains 12 sizing parameters, while the Folded VCOTA contains 33.

The datasets of the circuits also contain the required types and ranges of performances, which the model will use as guidance. The VCOTA contains the DC Gain (G_{DC}), Gain-Bandwidth Product (GBW), Bias Current (I_{DD}), and Phase Margin (PM). The folded VCOTA on the other hand, contains all the previous performance

P. H. M. Eid et al., *Efficient Analog Integrated Circuit Sizing with GenAI*,
SpringerBriefs in Computational Intelligence,
https://doi.org/10.1007/978-3-031-87105-4_4

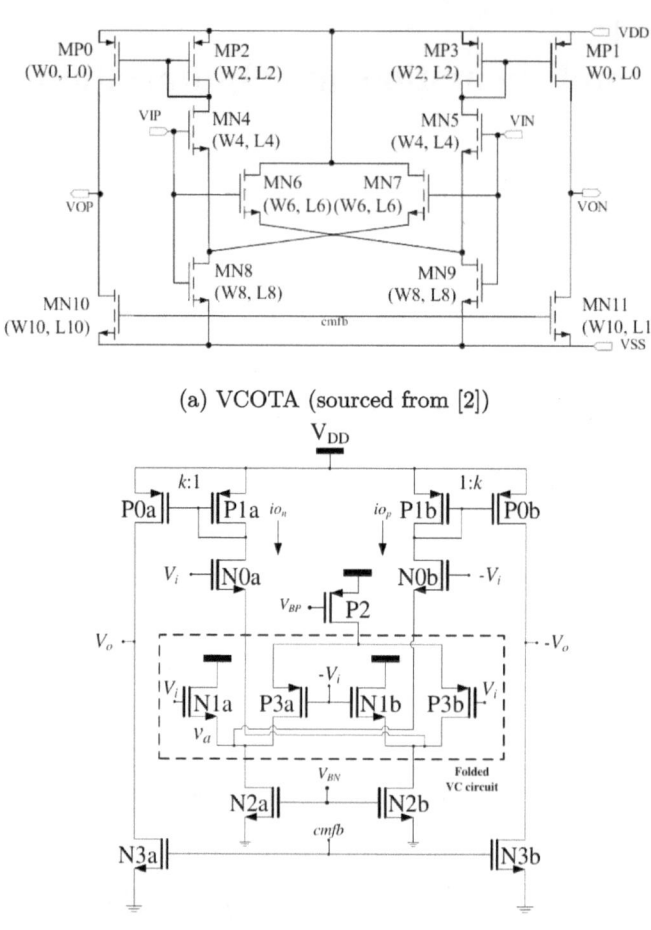

(a) VCOTA (sourced from [2])

(b) Folded VCOTA (sourced from [3])

Fig. 4.1 Circuits schematics

and includes also a Capacitance Load (C_{Load}) (the output impedance). This guidance parameters orients the denoising process to generate new sizing that produce circuits within the desired performance ranges. The different performance and load for each circuit are described in Table 4.1.

The VCOTA and Folded VCOTA, along with their corresponding performances, are extracted from datasets similar from the ones utilized in [3, 4], respectively. It's important to note that these datasets exclusively contain optimal or quasi-optimal designs for the sizing problem. In other words, the model is trained solely on optimal or nearly optimal solutions. The performance design space of the dataset can be seen the Fig. 4.2.

Table 4.1 Performance and load ranges

Circuit type	Range	G_{DC} [dB]	GBW [MHz]	I_{DD} [μA]	PM [°]	C_{Load} [fF]
VCOTA	Max	56.44	93.73	474.76	77.94	
	Min	44.73	37.40	243.84	50.14	
Folded VCOTA	Max	54.5	66.3	350	80.9	10000
	Min	45.0	30	30	60.0	500

As previously mentioned, the dataset is a noteworthy divergence between the proposed model and other applications of diffusion models. In contrast to common implementations that utilize a matrix (image) as the model input. This approach adopts a vector representation, a 12×1 vector for the VCOTA and 33×1 for the Folded VCOTA, with a size of 16661 and 8683 samples respectively. The ramifications of this dataset structure and the necessary adaptations have been previously mentioned in Chap. 3, but in this chapter its implementation will be discussed with more depth.

For training, optimizing, and evaluating the different networks and the diffusion model, the dataset was partitioned into train (80%), validation (10%), and testing (10%) sets, respectively.

4.1.2 Feature Engineering

Feature engineering involves applying data preprocessing methods and augmentation techniques to enhance datasets, ensuring better performance in machine learning models. This section highlights key approaches and strategies employed for this problem.

4.1.2.1 Data Preprocessing Methods

To ensure consistency and mitigate potential errors in training, the dataset underwent preprocessing methods. These methods are crucial due to the disparate dimensions of the variables involved in sizing and performance metrics. Parameters range from femto (e.g., C_{Load}) to mega (e.g.,GBW), and without them, the training of Artificial Neural Network (ANN) weights could be prone to errors. The first method used was a min-max normalization, where all parameters are scaled to the range of $[-1, 1]$, using the Eq. 4.1.

Fig. 4.2 Range of the
guidance part of the dataset

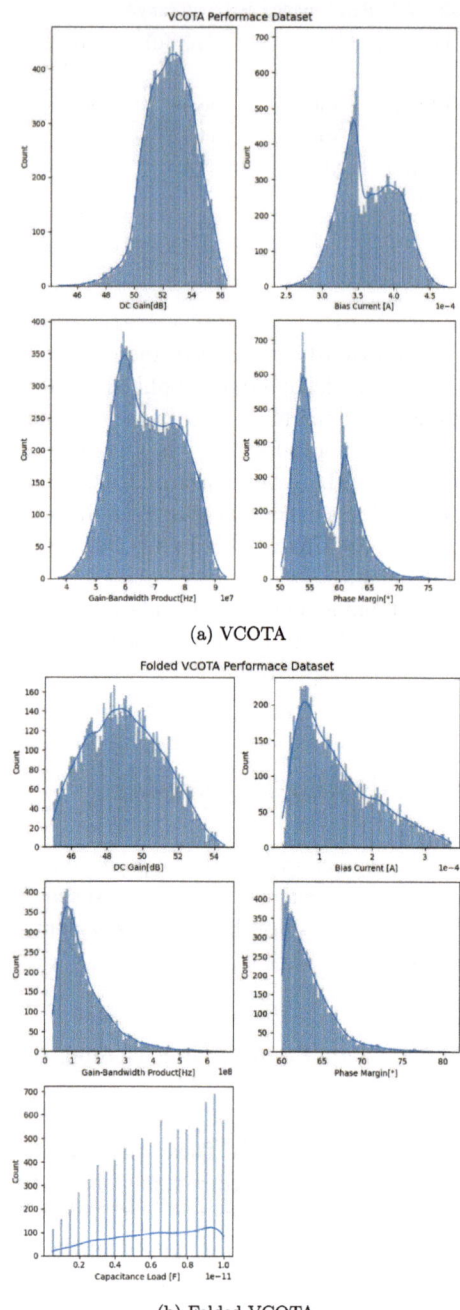

(a) VCOTA

(b) Folded VCOTA

$$X' = \left(\frac{X - X_{\min}}{X_{\max} - X_{\min}} \cdot 2 \right) - 1 \tag{4.1}$$

After testing this approach, standardization was also tested. This method, employing the Eq. 4.2, scales the variables to have a mean of 0 and a standard deviation of 1. The second preprocessing method, initially, yielded more consistent results, likely because the dataset was skewed and with outliers that could have led to disparities in normalization results. Standardization, being more robust to skewed dataset, resulted in improved consistency and prediction accuracy.

$$X' = \frac{X - \mu}{\sigma} \tag{4.2}$$

In addition to these methods, a log-transformation was applied before normalization and standardization. This transformation was initially tested with the sizing parameters, resulting in improved stability in predictions. This improvement is expected, as most of the sizing dataset is skewed, as illustrated by the example illustrated in Fig. 4.3 that plots the total width per finger (W10) of the VCOTA. Moreover, with this transformation the model is forced to generate only positive values for the sizing. Subsequently, the log-transformation was extended to include the performance parameters.

Ultimately, after the log-transformation, without a skewed dataset, the min-max normalization approach proved more effectively in ensuring consistent results across the datasets.

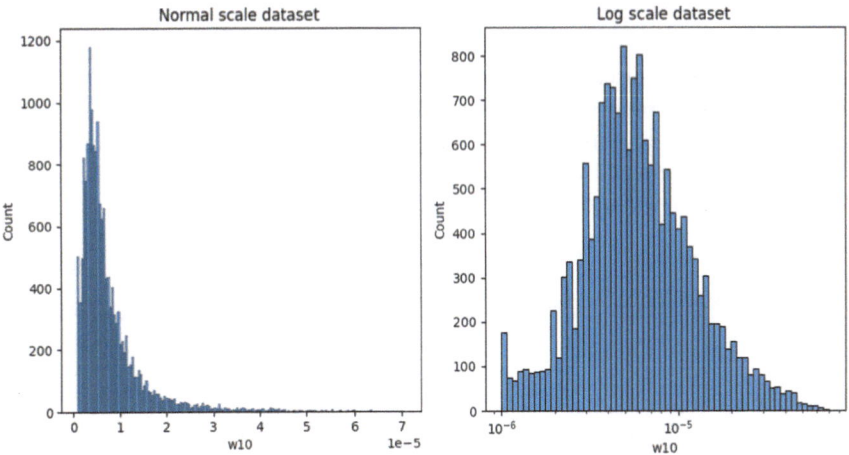

Fig. 4.3 Histogram of the VCOTA sizing parameter w_{10}

4.1.2.2   Feature Augmentation

Another feature engineering tool tested before feeding the data to the model was Polynomial Feature Augmentation. Initially, this augmentation was applied to both the sizing parameters and the performance parameters. However, transforming the sizing predictions back to their original form would result in discarding part of the learned knowledge from the model. Therefore, this augmentation was restricted to only the performance parameters. These new features enhanced the model's capacity to discern intricate relationships among different parameters and their polynomial counterparts, subsequently improving the model's convergence. This transformation can be seen in the Eq. 4.3.

$$(p_1, p_2 \ldots, p_n) \xrightarrow{\text{PF}} (p_1, p_2, \ldots, p_n, p_1^2, p_1 \cdot p_2, \ldots, p_1 \cdot p_n, p_2^2, p_2 \cdot p_3, \ldots, p_n^2)$$
(4.3)

With this transformation, the number of performance parameters increased from 4 and 5 to 14 and 20 for the VCOTA and Folded VCOTA, respectively. After this augmentation, the polynomial vectors are used to modulate the time embedding and then fed to the Denoising Diffusion Probabilistic Models (DDPM). It is worth noting that this transformation was not applied to the performance of the auxiliary ANN for the same reason, it would cause the discarding of the model's learned knowledge.

4.2   Supervised Learning Models

The first model to be discussed in this chapter is the Multi-Layer Perceptron (MLP) for the Supervised Learning (SL) approach. This MLP takes as input a log-transformed, min-max normalized polynomial of the performance parameters and predicts the circuit sizing. The model was firstly optimized with validation loss and the Optuna library. The optimal values obtained are shown in the table below (Table 4.2).

Table 4.2 Hyperparameter for the MLP in the supervised learning approach

Hyperparameters	Range	Supervised MLP optimal values	
		VCOTA	Folded VCOTA
Number of hidden layers	[2, 20]	13	17
Average number of neurons	[20, 5000]	1041	846
Batch size	[32, 500]	34	67
Learning rate	[1e–5, 1e–2]	4.1e–06	2.7e–05
Number of epochs	[1:200]	56 (34 min)	123 (27 min)

For this optimization, 20 trials were conducted, each with a different set of hyper-parameters with 1000 epochs. It is worth noting that an early stopping was also implemented for this ANN. It is important to highlight that the loss function and activation function remained consistent throughout each trial. The loss function used was Mean Square Error (MSE) Loss, while the activation function employed was Parametric Rectified Linear Unit (PReLU). Additionally, the Adam Optimizer was utilized with a weight decay of 1×10^{-7}, which acts as L2 regularization, contributing to the regularization of the model.

To further compare the proposed model with a state-of-the-art model, the ANN-1 from the paper [4] was recreated. This model was selected because, according to the paper, it produced the most flexible results and did not incorporate the inequality augmentation proposed in that work.

The optimized SL model demonstrated significantly more stability and accuracy. Consequently, the state-of-the-art model was trained solely for comparison purposes with the VCOTA dataset. Its results will be discussed in more detail in the next chapter.

4.3 VCOTA Dataset

This section focuses on a description of the implementation and optimization of the different configurations of the model with the VCOTA. The same steps were performed for the Folded VCOTA. The results and steps of this second optimization are described more succinctly in Sect. 4.4.

4.3.1 Multilayer Perceptron for Evaluation

In this first subsection, the focus is on the implementation of the Auxiliary ANN and its optimization process. As previously mentioned, the metric used for the initial evaluation of the different configurations of the DDPM is based on the relative error calculated from the performance of the sizing prediction. This is accomplished by passing the generated values through the auxiliary network and comparing them with the target performance.

The auxiliary network itself is a simple MLP, which reuses the same implementation as the MLP used in the SL approach. The optimization of this ANN uses the validation loss, and the Optuna [5] Python library. The parameters used in this optimization, along with their respective ranges and the optimal values found, are detailed in Table 4.3.

A total of 20 trials were conducted during this optimization, with each trial using a different set of hyperparameters and running for 1000 epochs. As previously mentioned, this model utilizes the same structure as the prior MLP, with identical loss functions, activation functions, weight decay, and an early stop.

Table 4.3 Hyperparameter of the auxiliary network

Hyperparameters	Range	Optimal values
Number of hidden layers	[2, 20]	9
Number of neurons	[20, 5000]	[3281, 110, 1139, 94, 3006, 1417, 1398, 535, 37]
Batch size	[32, 500]	119
Learning rate	[1e–6, 1e–4]	3.55e–05
Number of epochs	[1:500]	64 (12 min)

Upon evaluation with the test dataset, the optimized network demonstrated robust performance predictions, achieving an average mean relative error of less than 2% for the performance parameters.

4.3.2   Denoising Diffusion Probabilistic Models

This section focuses on the implementation and optimization of DDPM models for the VCOTA dataset.

4.3.2.1   Implementation of Different Processes

At the start of implementing the DDPM, the first objective was to create a noise schedule that defines the hyperparameters of the diffusion model, namely β, α, and $\bar{\alpha}$. The initial attempt involved implementing a linear schedule. However, testing revealed that noise was being introduced too rapidly into the dataset. Consequently, a cosine schedule was utilized in this project. Once the noise schedule was defined, the forward schedule was implemented. This process is straightforward leveraging the Eq. 2.4 and the re-parameterization trick, commonly employed in Variational Auto-encoder (illustrated in Eq. 4.4).

$$p(x_t|x_{t-1}) = \mathcal{N}(x_t; \mu_{x_{t-1}}, \sigma_{x_{t-1}}^2)$$
$$= \mu_{x_{t-1}} + \sigma_{x_{t-1}}\epsilon \quad \epsilon \in \mathcal{N}(1, 0) \tag{4.4}$$

This process introduces noise to the sizing dataset. The main goal is to produce enough noise to transform the original distributions of the dataset's features into a normal distribution, which is then used to train the model to learn the original distribution. The process of adding noise and the resulting normal distributions can be seen in Fig. 4.4.

Following the implementation of the forward process, the sampling process was implemented according to the pseudocode presented in Fig. 3.2. It is noteworthy

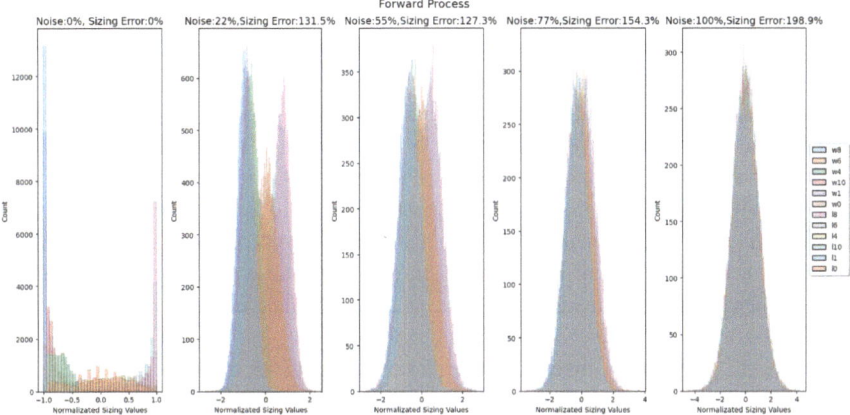

Fig. 4.4 Forward process for the VCOTA dataset

that, after all iterations of the sampling process, a manual clamping was applied to ensure that the sizing parameters remained within the specified range of minimum and maximum values of the technologies. This adjustment aimed to confine the predictions within the physical dimensions of the sizing components.

After implementing the sampling process, the reverse process was constructed based on the loop outline in Fig. 3.2. Subsequently, the Guidance mechanism was integrated into this loop, which proved relatively straightforward. During the reverse step, based on a random probability governed by the hyperparameter p (set to 0.1), the normalized polynomial performance vector is used to modulate the time embedding.

Various modulation techniques, as previously mentioned, were evaluated. Multiplication yielded superior results compared to simple addition, though the affine transformation demonstrated even better guidance. Based on this observation, affine transformation for subsequent implementations and optimizations was chosen.

Following the implementation of the guidance in the reverse process, the next step was incorporating it into the sampling process using the following equation:

$$\tilde{\epsilon} = (1 + w)\epsilon_\theta(x_t, c) - w\epsilon_\theta(x_t) \qquad (4.5)$$

Here, $\epsilon_\theta(x_t, c)$ represents the noise learned by the conditional network, while $\epsilon_\theta(x_t)$ pertains to the noise learned by the unconditional network. The hyperparameter w plays a crucial role in adjusting the strength of the guidance and in removing the unconditional knowledge from the sampling process. The value of w was later optimized for different configurations of the model.

4.3.2.2   Optimization of the DDPM

The DDPM includes several hyperparameters, such as the guidance weight, probability of the guidance mechanism, and the number of time steps. This section primarily focuses on two of these hyperparameters: the weight of the guidance mechanism and the number of time steps. To optimize these values, the various networks were utilized (whose implementations will be explored later in this chapter), the auxiliary network, and the validation dataset. To facilitate this step, a preliminary optimization was conducted with the different architectures. With these semi-optimized models these hyperparameters (weight and time step) were changed, and the results were recorded.

The first optimization discussed in this section involves a component of the forward step: the number of time steps. Altering this parameter does not change the overall amount of noise added in the forward step but rather affects the number of distinct steps at which noise is introduced into the dataset. This hyperparameter is crucial because it determines the extent of data degradation at each step and the number of steps the ANN needs to learn to denoise. Various values were tested to observe their impact on performance error, as shown in Fig. 4.5.

As can be observed from the graph, the optimal results are achieved within the range of ten to one hundred. This marks a notable divergence from the image's diffusion literature, where optimal results are generally attained with values in the range of hundreds or even thousands, with accuracy improving as the number of time steps increases. The discrepancy between the implementation and the literature may be attributed to the nature of the models or its smaller dataset size.

The second optimization involves the weight (w) of the guidance mechanism. This weight controls the balance between the conditional and unconditional components

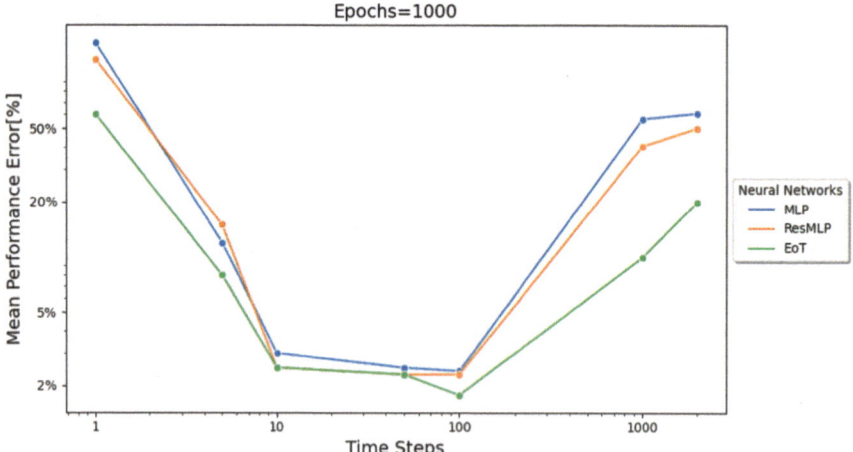

Fig. 4.5 Number of time step with the VCOTA dataset

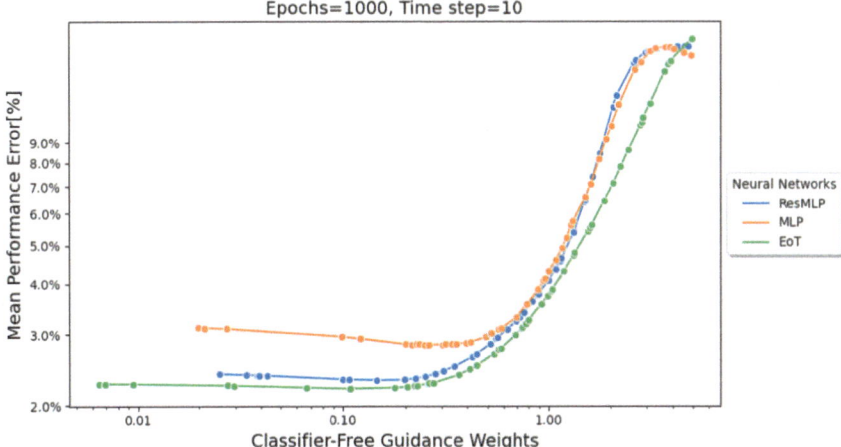

Fig. 4.6 Plot of weight values of the guidance mechanism with the VCOTA dataset

of the DDPM during the sampling process, as illustrated by Eq. 4.5. Increasing the weight value enhances the dependency on the conditional component of the model, effectively "subtracting" the influence of the unconditional component. Conversely, decreasing the weight value preserves the influence of the unconditional component, thereby enhancing the prediction's diversity. In the literature, values around $w = 0.3$ are commonly used to achieve an optimal balance.

This optimization step was conducted using the Optuna library [5]. The DDPM was trained with different networks and sampled with various weights, with the mean performance error observed. The tested values and results are shown in Fig. 4.6.

As can be observed in the graph, for all models, the values that yielded the best results were in the range of 0.1–0.3, which aligns with the values commonly used in the literature. It is worth noting this optimization was redone after the final optimization of the ANN, but the trend continued the same for this hyperparameter.

4.3.3   Neural Network Implementations

Following the construction of the DDPM, attention shifted to the ANN, focusing on its construction and optimization. The ANN within the diffusion model receives as input the noisy data, temporal information, and conditional information (when applicable). The latter two are modulated as previously mentioned. After modulation, the result is added to the sizing input. This combined vector is then fed into the hidden layers of the ANN, which outputs the predicted noise of each time step.

The final evaluation of the different ANNs will be discussed in depth in the following chapter with the real simulator. Since the auxiliary network produces preliminary results that can contain errors (approximately 2%), it cannot be entirely relied upon for performance results, especially for generated sizing close to the boundaries or outside the dataset. Nonetheless, these optimized models were able to achieve a performance error, as calculated by the auxiliary network, of around 2% for the VCOTA.

4.3.3.1 Multilayer Perceptron Implementation

The first model to be implemented was the MLP. Its implementation was straightforward by following the diagram in Fig. 3.8. The optimization was conducted by using the Optuna library, leveraging the auxiliary network, and validation datasets. By directly comparing the auxiliary network's performance predictions with target performance of the validation set, the errors for the various proposed hyperparameter configurations were calculated. Ten trials each employing a unique hyperparameter configuration selected by the Bayesian Optimization (BO) algorithm. Each trial run for 1000 epochs, and the resulting errors are visualized in the Fig. 4.7.

In this optimization, the mean performance error is calculated every 100 epochs, starting from epoch 1. The predicted sizing for the validation set is evaluated by the pre-trained auxiliary network to compute the performance based on these predictions, and consequently, the error between the predicted performance and the target performance. This relative error serves as the primary objective for BO.

After identifying the most optimal hyperparameter configurations, the network is trained for up to 5000 epochs, using three different time steps. The configuration that results in the smallest performance error is selected. Once these optimal configurations are determined, the trained model undergoes guidance weight optimization. In this process, the trained ANN is used in conjunction with Optuna and the auxiliary network to calculate the optimal weight. The range of values considered during

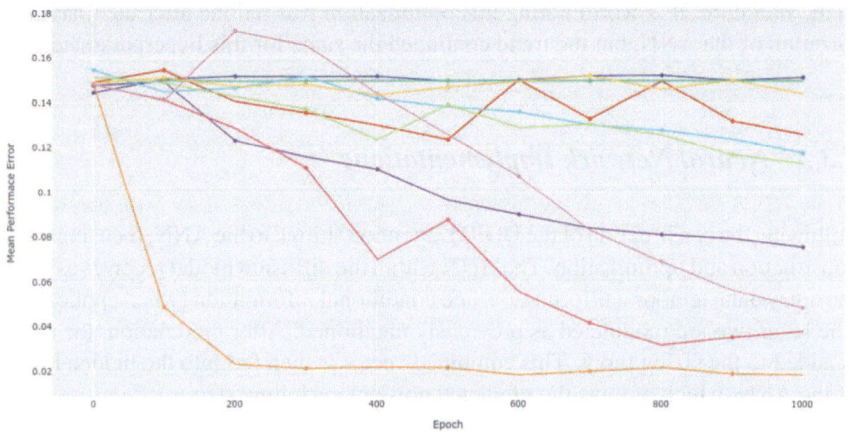

Fig. 4.7 Epoch graph for the different optuna trials for MLP with VCOTA dataset

Table 4.4   Multilayer Perceptron with VCOTA dataset

Hyperparameters	Range	Optimal values
Number of hidden layers	[2:30]	7
Number of neurons	[20:5000]	[467, 292, 758, 1115, 208, 1968, 507]
Loss type	[l1, mse]	l1
Batch size	[32:500]	175
Learning rate	[1e–5, 1e–2]	9.3e–4
Guidance weight	[0.0:5.0]	0.090
Time steps	[10, 50, 100]	10
Number of epochs	[1:5000]	2300 (24 min)

these optimization steps, the optimal values obtained, and the epochs used to train the model are detailed below (Table 4.4).

The goal of the reverse process of the DDPM is to learn how to recreate the original distribution of the input (sizing) data from a Gaussian distribution. As can be seen in Fig. 4.8, this model was able to produce results that closely follow the original distribution for the different sizing parameters.

4.3.3.2   Residual Multilayer Perceptron Implementation

After completing the optimization of the MLP, focus was given on implementing the Residual Multilayer Perceptron (ResMLP). This implementation is similar to the previous one and follows the structure shown in Fig. 3.8, with a key difference: the output of the layers in the first half of the network is added to the input of the layers in the second half.

The optimization also follows the same approach by using the auxiliary network, Optuna library, and validation set. As with the previous optimization, 10 trials were performed with 1000 epochs each. The results can be observed in Fig. 4.9.

By analyzing the optimization graphs in Figs. 4.7 and 4.9, it becomes apparent that the different configurations of the latter model seem more stable, whereas the configurations of the first model often get stuck at higher error percentages. This can be explained by the use of residual connections, which help the model retain information and not overfit.

It is worth noting that since the outputs of the initial layers are added to the latter layers, these latter layers must have the same size as the initial ones. Consequently, the number of layers for which Optuna needs to predict the number of neurons is at most half of the total number of layers. The table below shows the range and optimal values used and obtained in this optimization (Table 4.5).

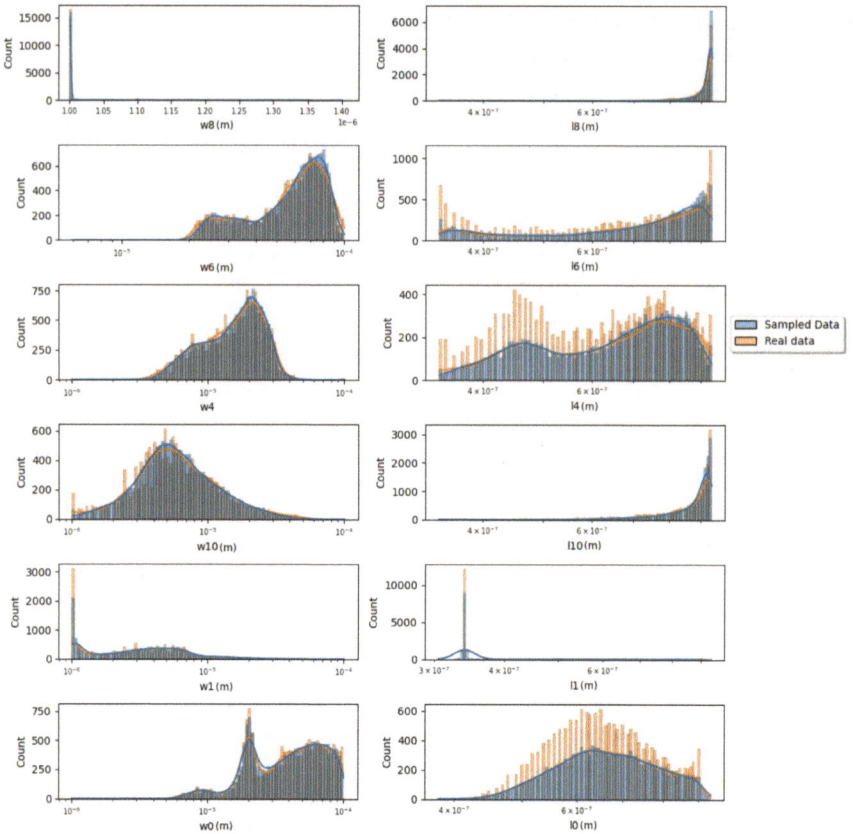

Fig. 4.8 Distributions of sample data and the real distribution for MLP with the VCOTA dataset

The distribution of the samples from this optimized model does not show significant visual differences compared to the distribution of the previously optimized model, that is observed in Fig. 4.8.

4.3.3.3   Encoder-Only Transformer Implementation

The final model implemented was the Encoder-only Transformer (EoT). This model builds upon the ResMLP architecture by incorporating a Multi-head Self-Attention block before each hidden layer. The optimization process follows the same approach as before, consisting of 10 trials, each with 1000 epochs. And by using the Optuna library the epochs of Fig. 4.10 were obtained.

By comparing the graphs produced from the optimization of all the different ANN, it can be observed that this model appears to be the most stable, with all configurations

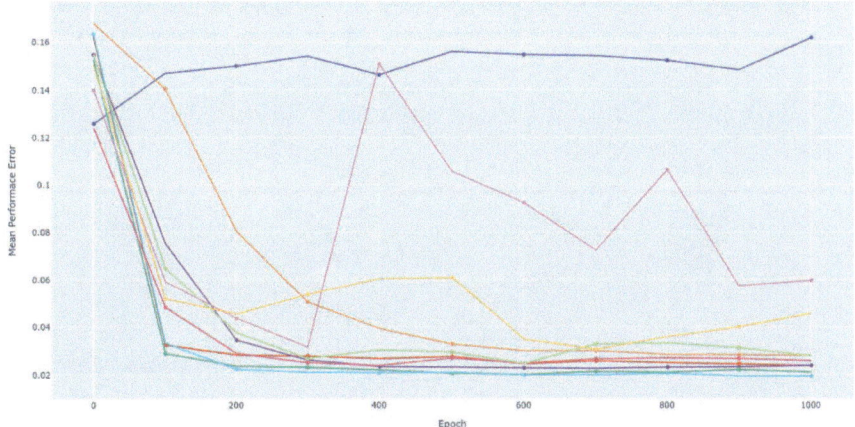

Fig. 4.9 Epoch graph for the different optuna trials for ResMLP with the VCOTA dataset

Table 4.5 Hyperparameter of the residual multilayer Perceptron with the VCOTA dataset

Hyperparameters	Range	Optimal values
Number of hidden layers	[2, 30]	27
Number of neurons	[20, 5000]	[1765, 1794, 209, 1780, 203, 229, 696, 731, 1429,1758, 218, 381, 1432, 315, 811, (...)]
Batch size	[32, 500]	206
Loss type	[l1, mse]	l1
Learning rate	[1e–5, 1e–2]	8.27e–05
Guidance weight	[0.0, 5.0]	0.39
Noise steps	[10, 50, 100]	10
Number of epochs	[1:5000]	3400 (56 min)

yielding predictions with a small error value. The range of values tested and the optimal values for this network are presented in the table below (Table 4.6).

4.4 Folded VCOTA Dataset

After optimizing the model for the VCOTA, attention was shifted to the larger circuit, the Folded VCOTA. This circuit is more intricate, featuring almost three times the number of sizing parameters and incorporating a C_{Load}. The C_{Load}, combined with performance parameters, guides the denoising step. Notably, this parameter serves as an input for both the DDPM and the auxiliary network. With this new dataset that contains almost half the data points of the previous dataset, the same feature

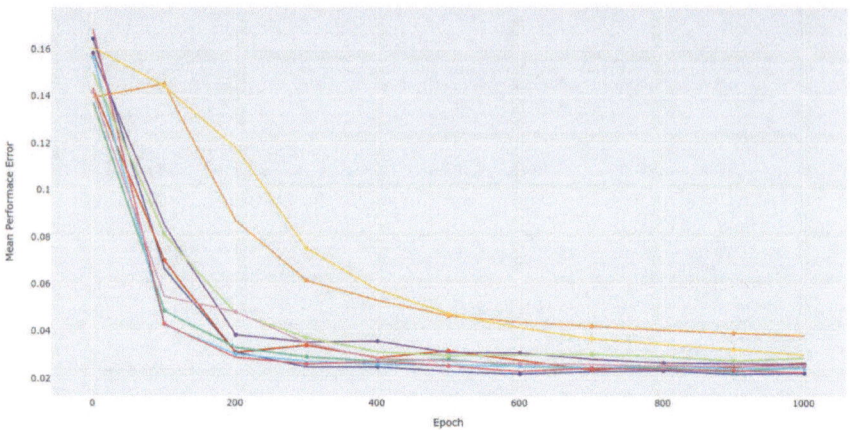

Fig. 4.10 Epoch graph for the different optuna trials for EoT with the VCOTA dataset

Table 4.6 Hyperparameter of the encoder-only transformer with the VCOTA dataset

Hyperparameters	Range	Optimal values
Number of hidden layers	[2, 30]	21
Number of neurons	[20, 5000]	[1082, 529, 798, 536, 1958, 490, 518, 206, 972, 201, 1743, (...)]
Number of attention heads	[1, 32]	[3, 2, 13, 7, 1, 29, 1, 1, 8, 30, 1, 3, 3, 3, 2, 25, 1, 1, 27, 11, 25, 1]
Batch size	[32, 500]	177
Loss type	[l1, mse]	l1
Learning rate	[1e–5, 1e–2]	3.18e–05
Guidance weight	[0.0, 5.0]	0.18
Noise steps	[10, 50, 100]	10
Number of epochs	[1:5000]	4300 (84 min)

engineering techniques were applied, specifically polynomial transformation, logarithmic transformation, and min-max normalization.

Following the preprocessing of the dataset, the focus turned to the DDPM. The model, initially developed for the previous dataset, required only small adjustments to accommodate the different input and guidance vector size. The forward process for this dataset is illustrated in Fig. 4.11.

After adapting the DDPM, focus was given on its optimization. For this dataset, similar trends were observed in the hyperparameters, noise steps and guidance weight, as in the previous dataset, as shown in Figs. 4.5 and 4.6. It is important to note that because this dataset includes the number of fingers, these parameters are rounded to the nearest integer after each sampling.

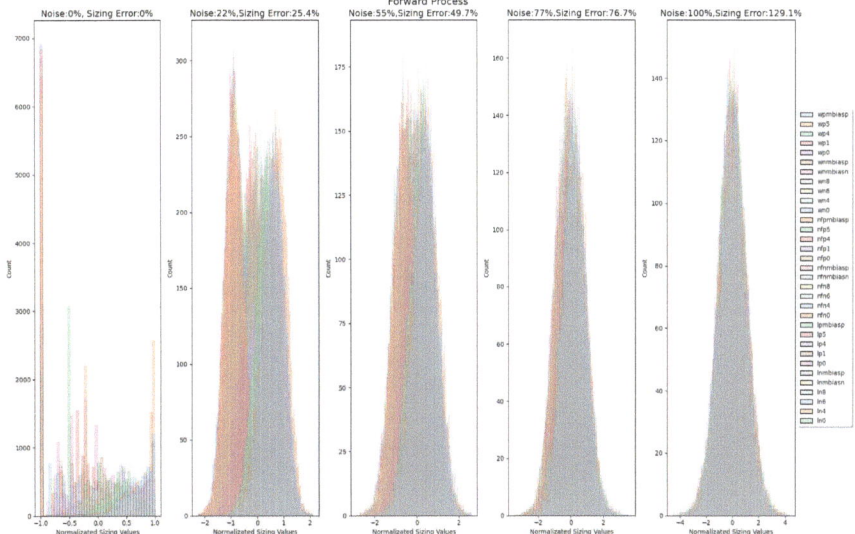

Fig. 4.11 Forward Process with the Folded VCOTA dataset

4.4.1 Neural Network Optimization

With the dataset prepared and the DDPM adapted, the focus was shifted to optimizing the various ANN models. The auxiliary network was firstly optimized, following the same steps used for the previous dataset. Subsequently, the other ANN models were optimized. This process again utilized the validation portion of the dataset, the optimized auxiliary network, and the Optuna library. For this optimization, the number of trials was increased from 10 to 20, as the MLP did not find a good solution within just 10 trials. The various graphs produced during this optimization process are shown in the Fig. 4.12. These graphs show the mean error values after 1000 epochs for different hyperparameter configurations of various ANN models. As observed, all models faced greater difficulty achieving low mean errors compared to the previous dataset. The MLP only found a configuration with an error below 20% by its 13th trial. Although the other ANN models identified good configurations more quickly, some configurations resulted in errors exceeding 30%. Such high errors were not observed in the previous ANN configurations for the VCOTA.

The table below displays the ranges and optimal values obtained for the different ANN models optimized in this process (Table 4.7).

After the models were optimized, each one was sampled, and it was found that all were able to reproduce the original distribution, as an example shown in Fig. 4.13 for the EoT. It is worth mentioning that for this dataset, a mean relative error of around 3% on the test set was achieved using the auxiliary network across the different ANN models.

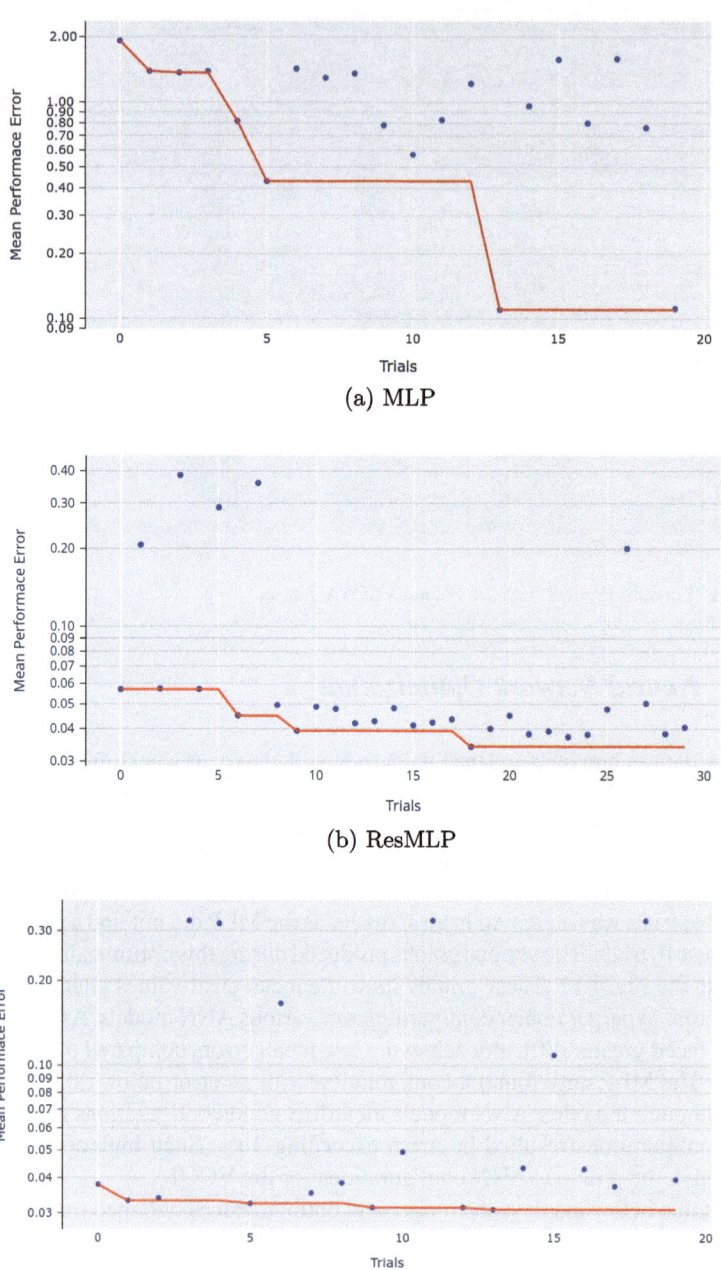

Fig. 4.12 Graph of the performance values for the different optuna trials with folded VCOTA dataset

Table 4.7 Hyperparameter for the folded VCOTA dataset

Hyperparameters	Range	Optimal values			
		Auxiliary ANN	MLP	ResMLP	EoT
Number of hidden layers	[2, 30]	9	3	13	12
Average number of neurons	[20, 5000]	1272	1268	917	371
Average number of attention heads	[1, 32]	–	–	–	6
Batch size	[32, 500]	252	117	166	155
Loss type	[l1, mse]	l1	l1	l1	mse
Learning rate	[1e–5, 1e–2]	4.5e–05	1.6e–3	5.4e–4	8.78e–05
Guidance weight	[0.0, 5.0]	–	0.179	0.313	0.1613
Noise steps	[10, 50, 100]	–	10	10	10
Number of epochs	[1:5000]	130 (4 min)	3600 (14 min)	4300 (33 min)	4600 (48 min)

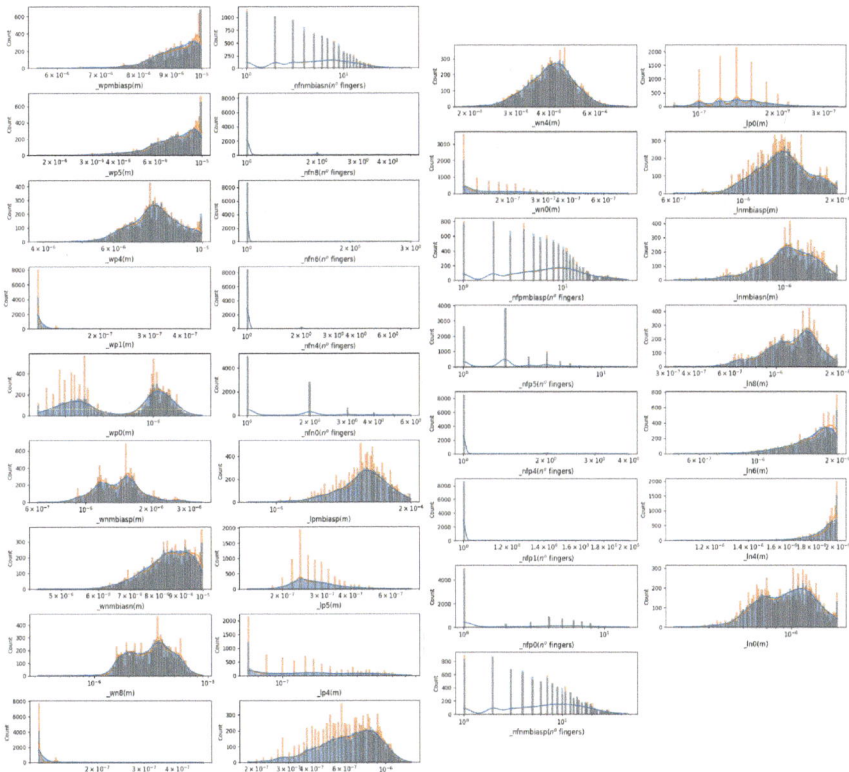

Fig. 4.13 Distributions of sample data and the real distribution for the EoT with the folded VCOTA dataset

4.5   Conclusion

In this chapter, the various implementations and optimizations of the proposed architectures were discussed. The chapter began with an analysis of the dataset and the implemented feature engineering techniques. Following that, the implementation of the SL models, auxiliary network, DDPM, and denoising ANN with the VCOTA dataset was explored. The implementation of the folded VCOTA was then addressed, with all the denoising ANN in both datasets being able to recreate the initial distributions from a white Gaussian noise for both datasets.

References

1. Paszke, A., et al.: PyTorch: an imperative style, high-performance deep learning library (2019). arXiv:1912.01703
2. Póvoa, R., et al.: Single-stage OTA biased by voltage-combiners with enhanced performance using current starving. IEEE Trans. Circuits Syst. II: Express Briefs **65**(11), 1599–1603 (2018). https://doi.org/10.1109/TCSII.2017.2777533
3. Povoa, R., et al.: A folded voltage-combiners biased amplifier for low voltage and high energy-efficiency applications. IEEE Trans. Circuits Syst. II: Express Briefs **67**(2), 230–234 (2020). ISSN: 1558-3791. https://doi.org/10.1109/TCSII.2019.2913083
4. Lourenco, N., et al.: On the exploration of promising analog IC designs via artificial neural networks. In: 2018 15th International Conference on Synthesis, Modeling, Analysis and Simulation Methods and Applications to Circuit Design (SMACD). IEEE (2018). https://doi.org/10.1109/SMACD.2018.8434896
5. Akiba, T., et al.: Optuna: a next-generation hyperparameter optimization framework (2019). arXiv:1907.10902

Chapter 5
Experimental Results

5.1 VCOTA Dataset

Using the Voltage Combiners biased Operational Transconductance Amplifier (VCOTA) dataset, the model was tested in two primary ways. First, the test portion of the data was predicted, and the predicted sizes were simulated. Second, four arbitrary performance points were selected, and the model's ability to predict these points was evaluated using 100 samples for each. The accuracy and difficulty of making the predictions for these arbitrary values were then assessed.

5.1.1 Test Predictions

The test dataset of the VCOTA, as previously mentioned, contains 1667 (10%) random chosen points, which were not previously used to train or model optimization. The points selected for this evaluation are illustrated in the Fig. 5.1.

To assess model accuracy on the unseen data, the predictions of both generative and Supervised Learning (SL) models were evaluated. The error rates for each model are presented in Table 5.1. By comparing these results, the relative effectiveness of this approach can be gauged against a more traditional approach.

As shown in Table 5.1, the three Denoising Diffusion Probabilistic Models (DDPM) models achieved an error rate of around 5%, while the SL had an error rate of approximately 7%. The model from the paper [1] had the worst performance with almost 9% error. Notably, the ResMLP and EoT models were able to make predictions with errors below 10% for all performance parameters. Overall, the generative approaches consistently outperformed the SL approach, demonstrating lower error rates across all evaluated performance parameters. Furthermore, this benchmark model surpassed the state-of-the-art.

P. H. M. Eid et al., *Efficient Analog Integrated Circuit Sizing with GenAI*, SpringerBriefs in Computational Intelligence, https://doi.org/10.1007/978-3-031-87105-4_5

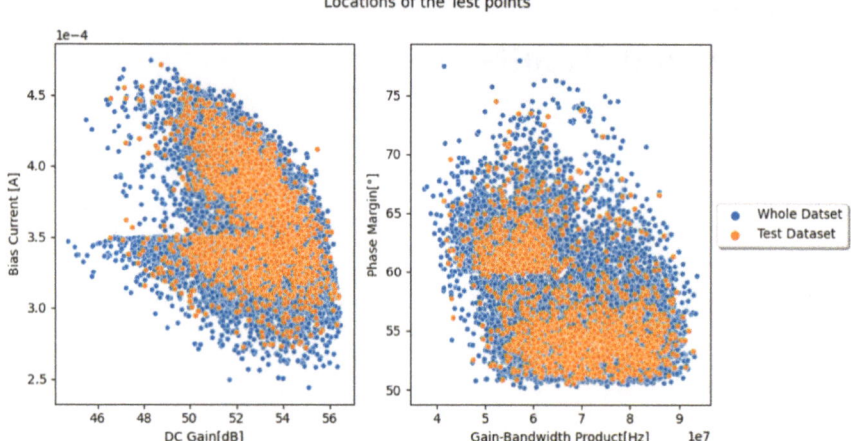

Fig. 5.1 VCOTA dataset performance points selected for testing

Table 5.1 Simulation test error of the VCOTA dataset

Performances	Median relative error (%)				
	MLP	ResMLP	EoT	SL	SL [1]
Gain-bandwidth product (GBW)	6.02	5.83	5.93	8.37	10.37
DC Gain (G_{DC})	0.88	0.90	0.77	1.43	4.75
Bias Current (I_{DD})	10.14	9.82	9.97	12.47	14.57
Phase Margin (PM)	4.50	4.76	4.9	6.46	5.38
Average	5.39	5.33	5.39	7.19	8.77

It is worth noting that for all models, including the SLs approaches, the percentage of points falling outside the saturation zone was below 10%. The MLP and SL approach exhibited slightly higher percentages, around 7%, while the other two models had percentages close to 4%.

5.1.2 Target Predictions

For this second test, four different performance points were selected. Their values can be observed in Fig. 5.2. Targets 0, 1, and 2 were taken directly from the paper [1] to facilitate comparison with a state-of-the-art model. Targets 0 and 3 represent relatively simple targets within the dataset, with Target 0 being closer to the dataset boundaries. In contrast, Target 1 is a challenging target located far outside the dataset. For Target 2, according to the paper [1], "there is no possible circuit sizing,

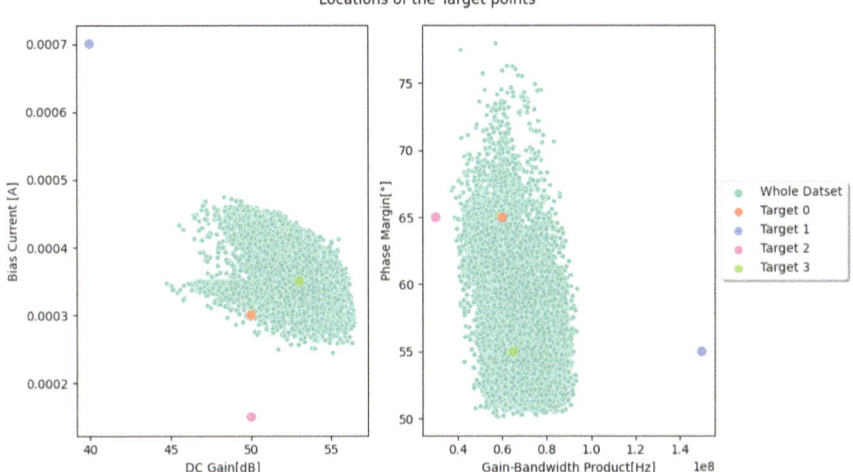

Fig. 5.2 VCOTA dataset performance points selected for inference

that we know of, that could meet this specification." This point was intentionally chosen to further test the generative capabilities of these models.

To fully leverage the generative capabilities of the diffusion models, a matrix of 400 elements was created, with each target consisting of 100 points. This matrix was used to guide the different DDPM models, and the generated sizes were then passed through the simulator to obtain real performance values. In contrast, due to its deterministic nature, for the SL model, only a matrix with the 4 points was utilized. As previously mentioned, for this test, the Artificial Neural Network (ANN) from the paper [1] was also used to further compare the diffusion models with a state-of-the-art approach.

After the performance data was obtained from the simulator, the error for each of the 400 points was calculated. In addition to identifying the point with the lowest mean error, the closest point (cp), the stochastic nature of the generative models was leveraged to calculate the closest value (cv) for all performance metrics for each target. In this table, the different generated performance values were compared, and the Figure of Merit (FoM) of the generated circuits was calculated for both datasets using Eq. 5.1.

$$FoM = \frac{\left(\frac{g_{bw}}{1000000}\right) \cdot 6}{i_{dd} \cdot 1000} \tag{5.1}$$

Except for Target 2 with the MLP, the generative models consistently produced points closer to the target values compared to the SL and achieving a higher FoM with all exceeding 1000. For the simpler targets, Target 0 and Target 3, the generative models delivered errors below 5%, with the MLP yielding the most accurate results.

In contrast, neither the SL approach nor the MLP achieved results below 20% for Target 1. In this target the EoT model performed the best with an error under 10%.

The Target 2 posed the most significant challenge, with none of the models producing acceptable results. With MLP consistently generated hallucinations for all 100 points, and neither the SL model nor the MLP managed to produce any points within the saturation zone. The other models also fell short, with errors exceeding 30% for all generated points.

Overall, when examining the cv of the targets, the MLP consistently delivered values closer to the targets than the other generative models. Nevertheless, all generative models provided reasonably close values, although the ResMLP produced the most distant values. In summary, generative approaches outperformed the SL approach in terms of accuracy and generalizability. Even when not achieving high precision on the different points, when taking the performance values separately they were able to generate values that were relatively close to the targets even for the performance values very outside the dataset.

Comparing the generative approaches, the MLP produced slightly more accurate points for targets within the dataset and showed closer cv values. However, it tended to hallucinate for targets further outside the dataset, yielding points that were even worse than those produced by the SL approach. In contrast, the other two generative approaches were less prone to hallucinations, with the EoT producing more consistent results.

When observing the state-of-the-art model, SL [1], it can be seen that the model does not appear to be accurate or generalize well to different targets. Even for the simple target 0, the model obtained an error greater than 50%. The decent results produced in the paper can be attributed to the sampling technique used, which involved producing 100 samples with different performance targets, allowing up to 15% deviation from the targets. This strategy could make sense for models with data augmentation since the data augmentation used this deviation, but for ANN-1 which did not use augmentation, this approach seems incompatible with the deterministic nature of SL approaches. Instead of testing the accuracy or generalizability of the model, this approach seems to attempt to emulate the generativity of models that were not implemented in the paper. As observed in Table 5.2, without this flawed sampling predictions, the model is not able to generate good target predictions.

It is worth noting that even for the models with augmentation, the wide range of values tested for each target with this sampling strategy seems to demonstrate the lack of accuracy of the models implemented on the paper. For example, the 15% deviation for G_{DC} of target 0 would use the ranges from 42, 5 to 50. This means the model uses a range larger than half of the entire dataset itself, including points outside the dataset values, in order to size even the simplest targets. Moreover, the model selects the closest performance points to the target without considering the actual performance metrics used to predict the sizing.

Table 5.2 Simulation target points of the VCOTA dataset

Targets	ANN	cp/cv	Performance values					Mean error (%)	Best FoM
			G_{DC}	I_{DD}	GBW	PM	FoM		
Target 0	–	–	**50**	**300e–6**	**60e6**	**65**	–	–	–
	SL	–	49.615	253.37e–6	40.612e6	75.87	961.71	**16.33**	961.71
	SL [1]	–	50.676	653.44e–6	104.66e6	35.80	961.02	**59.63**	961.02
	MLP	cp	49.29	315.97e–6	60.697e6	64.75	1152.6	**1.82**	1188.8
		cv	49.88	299.52e–6	60.39e6	64.83	–	–	
	ResMLP	cp	49.75	326.52e–6	62.41e6	63.806	1146.9	**3.81**	1181.0
		cv	50.0	299.83e–6	60.997e6	63.80	–	–	
	EoT	cp	49.78	321.92e–6	61.460e6	65.25	1145.5	**2.64**	1183.1
		cv	50.0	300.20e–6	59.85e6	65.25	–	–	
Target 1	–	–	**40**	**700e–6**	**150e6**	**55**	–	–	–
	SL	–	46.129	462.20e–6	85.07e6	59.37	1104.40	**25.13**	1104.40
	SL [1]	–	40.137	684.61e–6	35.611e6	33.66	312.09	**29.40**	312.09
	MLP	cp	39.78	141.43e–6	137.88e6	55.53	5849.52	**22.35**	5849.52
		cv	39.78	698.26e–6	151.29e6	55.176	–	–	
	ResMLP	cp	44.49	593.12e–6	113.09e6	53.713	1155.1	**13.36**	1165.3
		cv	44.214	610.86e–6	151.56e6	54.997	–	–	
	EoT	cp	42.80	683.53e–6	120.41e6	53.74	1056.9	**7.84**	1136.8
		cv	39.646	683.53e–6	146.46e6	55.222	–	–	
Target 2	–	–	**50**	**150e–6**	**30e6**	**65**	–	–	–
	SL	–	51.450	231.76e–6	52.108e6	74.17	1349.30	**36.6**	1349.30
	SL [1]	–	47.835	318.23e–6	24.63e6	31.37	464.45	**46.53**	464.45
	MLP	cp	38.507	154.98e–6	2.64e6	90.460	102.25	**39.16**	135.6
		cv	49.887	150.13e–6	28.16e6	64.837	–	–	
	ResMLP	cp	48.33	266.83e–6	41.606e6	60.53	935.54	**31.69**	1041.1
		cv	50.0	266.83e–6	30.98e6	63.80	–	–	
	EoT	cp	49.00	271.85e–6	42.38e6	54.64	935.53	**35.11**	1026.4
		cv	50.0	157.54e–6	29.589e6	65.25	–	–	
Target 3	–	–	**53**	**350e–6**	**65e6**	**55**	–	–	–
	SL	–	52.748	376.96e–6	66.958e6	51.885	1065.76	**4.21**	1065.76
	SL [1]	–	52.59	412.14e–6	73.40e6	50.44	1068.60	**9.94**	1068.60
	MLP	cp	52.79	353.38e–6	64.63e6	55.217	1094.1	**0.575**	1297.5
		cv	52.942	349.77e–6	64.971e6	55.176	–	–	
	ResMLP	cp	51.11	352.95e–6	62.19e6	53.7	1057.20	**2.77**	1212.3
		cv	53.042	350.21e–6	65.68e6	54.997	–	–	
	EoT	cp	51.11	348.18e–6	64.39e6	52.57	1109.62	**2.36**	1230.4
		cv	53.186	349.85e–6	64.893e6	54.769	–	–	

5.2   Folded VCOTA Dataset

After the evaluation of the VCOTA was completed, the folded VCOTA was assessed. As mentioned earlier, this dataset presents a greater challenge, containing nearly three times the number of sizing parameters compared to the previous one, yet with almost half the number of data points. The same two tests as before were conducted: the evaluation of the model predictions on the test set and the assessment of the inference capabilities of the models.

Additionally, a third evaluation method, referred to as the context-independent evaluation, was introduced. This involved the removal of a specific Capacitance Load (C_{Load}) value from the dataset, and the model was retrained using all available data except those with the excluded C_{Load}. The models were then tested using the previously excluded points to assess their ability to predict and adapt to previously unseen conditions. This approach allowed the performance of the models on this unseen load to be evaluated and further assessed the generative nature of this approach.

5.2.1 Test Predictions

As with the previous dataset, the optimized models were first evaluated using the Folded VCOTA with the test portion of the dataset. In this phase, 868 randomly selected points were used, which can be observed in Fig. 5.3.

By generating and predicting these points with the different models, the test sizing was obtained, and the simulator was then used to produce the predicted performances. These predicted performances were compared with the target performance, yielding the results presented in Table 5.3.

As can be observed by comparing these results with those from the previous dataset, all models had more difficulty predicting performance for this dataset compared to the previous one. The EoT and ResMLP models produced the best results, with the former achieving all performance with an error below 10%. The SL model had the worst results, with predictions errors exceeding 20% for the GBW.

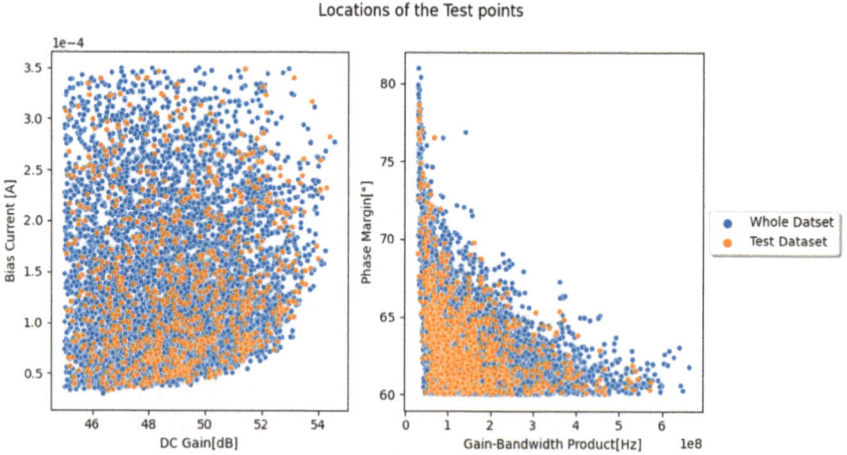

Fig. 5.3 Folded VCOTA dataset performance points selected for testing

Table 5.3 Simulation test error of the folded VCOTA dataset

Performances	Median relative error (%)			
	MLP	ResMLP	EoT	SL
GBW	12.29	10.92	9.74	20.66
G_{DC}	4.5	2.88	2.7	6.87
I_{DD}	10.97	7.75	7.28	9.87
PM	8.22	6.24	6.49	7.74
Average	9.01	7.01	6.54	11.30

In summary, all generative approaches outperformed the SL model. It is worth noting that all models also faced greater difficulty predicting points within the saturation zone, with more than 30% of all predicted points falling outside of saturation. The SL and MLP models had the highest number of these points, with more than 50% outside this zone.

Differently than the VCOTA circuit, that has the only dimensions predicted for the MOSFET devices with the total width and total length. This circuit, the folded VCOTA, the width being predicted is the width per finger, number of fingers and total length (considering that total width = width×finger and number of fingers). This is a more delicate prediction, as in the VCOTA the total width would present a "continuous" nature (i.e., at least in terms of manufacturing grid, e.g., 5 nm). In this folded VCOTA example each change in the number of fingers (i.e., integer only) results in at least an increment of width per finger (e.g., hundreds of nanometers or units of micrometers) in the total width of the devices. Naturally, more devices will fall for undesired overdrive voltages when the prediction of the number of fingers is not accurate.

This is possible to observe by considering the equation of the drain current for the MOSFET during saturation, as its direct dependence of the total width. Assuming that the mobility μ_n, oxide capacitance C_{ox} and threshold voltage V_{TH} are technology dependent, having a MOSFET device on a branch with a current I_{DD} imposed on it, for a fixed $(\frac{W}{L})$, only V_{GS} will be allowed to vary. By varying V_{GS} significantly, the overdrive voltage V_{OD} ($V_{OD} = V_{GS} - V_{TH}$) may increase or decrease, which consequently affects the fulfillment of the saturation margin or not ($V_{SAT} = V_{DS} - V_{OD}$). This is, if total Width is incremented by a width per finger multiple, VGS will have to decrease in order to keep the same I_{DD}, eventually even turning off the channel (i.e., $V_{GS} < V_{TH}$). If the total Width is decrease by a width per finger multiple, V_{GS} will have to increase in order to keep the same I_{DD}, eventually breaking the saturation condition (i.e., $V_{DS} < V_{GS} - V_{TH}$), or breaking the V_{SAT} specification ($V_{DS} - V_{OD} < V_{SAT_{spec}}$).

$$i_{dd} = \frac{1}{2}\mu_n \cdot C_{ox} \left(\frac{W}{L}\right) (V_{GS} - V_{TH})^2 \qquad (5.2)$$

5.2.2   Target Predictions

After the models were evaluated with test predictions, their generative inference capabilities were assessed. For this second evaluation, four arbitrary points were selected. To choose these points, an arbitrary value for the C_{Load} was first set, centered around the mode of this parameter (9500fF). Using this value, the targets were then defined as follows:

- **Target 0**: Located at the mean value of all performance parameters.
- **Target 1**, **Target 2** and **Target 3**: Positioned one, two, and three standard deviation away from Target 0, respectively.

As can be observed in Fig. 5.4, target 0 is located at the center of the dataset. Point 1 is situated at the boundaries of the dataset, while points 2 and 3 are located outside the dataset, with point 3 being the more difficult target. After selecting the four target points, predictions were made for each target with 100 generated samples using the DDPMs, and 4 prediction points were made with the SL. After processing the generated and predicted sizing through the simulator, the results were compared with the target values. The results, presented in Table 5.4, are divided into cv and cp categories, similar to the results from the previous dataset.

For targets 0 and 1, the generative models had the least difficulty predicting the performance values, with all models achieving errors smaller than 5%. Interestingly, the MLP achieved the best results, with less than 1% error for both targets. Surprisingly, the MLP and ResMLP were able to predict target 1 better than the mean target. The EoT also performed well, predicting all targets with an error below 10%. While the MLP had the best results for targets within the dataset, it again hallucinated for

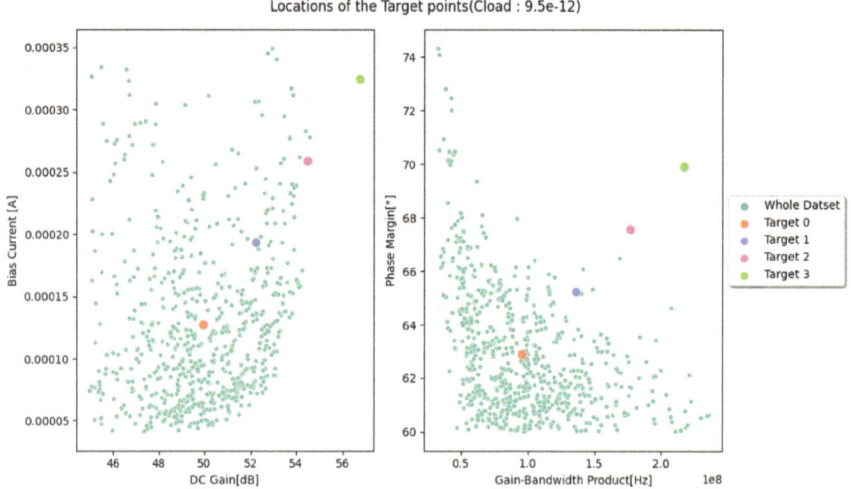

Fig. 5.4 Folded VCOTA dataset performance points selected for inference test

Table 5.4 Simulation target points of the folded VCOTA dataset

Targets	ANN	cp/cv	Performance values					Error (%)	Best FoM
			G_{DC}	I_{DD}	GBW	PM	FoM		
Target 0	–	–	**49.97**	**127.11e–6**	**95.88e6**	**62.89**	–	–	–
	SL	–	46.018	144.59e–6	84.37e6	71.63	3501.4	**11.88**	3501.4
	MLP	cp	48.92	127.18e–6	94.659e6	62.62	4465.7	**0.959**	5345.6
		cv	49.983	127.09e–6	95.92e6	62.88	–	–	
	ResMLP	cp	49.059	126.56e–6	90.12e6	64.96	4272.6	**2.89**	4327.8
		cv	49.988	127.03e–6	95.32e6	64.97	–	–	
	EoT	cp	48.99	124.57e–6	90.86e6	64.99	4376.3	**3.13**	5694.7
		cv	49.973	127.21e–6	95.82e6	63.93	–	–	
Target 1	–	–	**52.22**	**192.84e–6**	**136.48e6**	**65.22**	–	–	–
	SL	–	53.597	143.10e–6	47.95e6	75.448	2010.4	**27.23**	2010.4
	MLP	cp	51.74	194.10e–6	139.19e6	65.13	4302.7	**0.94**	4302.7
		cv	52.219	194.10e–6	134.98e6	65.193	–	–	
	ResMLP	cp	51.50	193.57e–6	138.41e6	65.71	4290.5	**0.98**	4297.7
		cv	52.193	193.57e6	134.90e–6	64.965	–	–	
	EoT	cp	49.20	194.19e–6	138.34e6	68.70	4274.38	**3.30**	4316.6
		cv	52.2	192.17e–6	138.34e6	65.19	–	–	
Target 2	–	–	**54.49**	**258.58e–6**	**177.08e6**	**67.55**	–	–	–
	SL	–	52.092	187.56e–6	75.68e6	78.057	2421.2	**26.16**	2421.2
	MLP	cp	35.22	320.69e–6	138.20e6	62.030	2585.5	**22.38**	2585.5
		cv	54.426	257.20e–6	169.38e6	67.596	–	–	
	ResMLP	cp	52.76	266.77e–6	147.08e6	69.823	3308.0	**6.656**	3308.0
		cv	54.072	259.31e–6	177.51e6	67.559	–	–	
	EoT	cp	52.90	259.27e–6	130.64e6	73.13	3023.3	**9.41**	3135.5
		cv	49.053	256.74e–6	168.37e6	69.442	–	–	
Target 3	–	–	**56.75**	**324.32e–6**	**217.68e6**	**69.89**	–	–	–
	SL	–	51.19	216.05e–6	92.48e6	77.78	2568.4	**27.99**	2568.4
	MLP	cp	28.22	78.97e–6	5.81e6	13.512	441.44	**75.98**	441.44
		cv	54.42	326.78e–6	218.63e6	69.95	–	–	
	ResMLP	cp	49.99	404.54e–6	212.53e6	71.754	3152.2	**10.42**	3152.2
		cv	54.072	322.76e–6	218.52e6	69.89	–	–	
	EoT	cp	52.18	341.10e–6	188.86e6	69.81	3322.0	**6.65**	3322.0
		cv	54.431	325.31e–6	218.36e6	69.883	–	–	

targets outside the dataset, failing to find a single point in saturation for target 3. The SL obtained the worst results (excluding the MLP with the target 3), consistently producing errors greater than 10% even for targets within the dataset and produced also sizing with a smaller FoM.

Taking a look at the cv, the generative models were able to produce values close to the target performance values, with most of the predicted values deviating by less than 1% from the target performance. In summary, the generative models outperformed the more traditional approach, achieving greater accuracy and better generative capabilities for both this dataset and the previous one.

5.2.3 Context-Independent Evaluation

The final evaluation of the generative models with this second dataset is related to the C_{Load} parameter. A context-independent evaluation was conducted by constructing the test set of the dataset using only instances with the same values of this parameter, while excluding these values from the training dataset. The arbitrarily chosen value (500 fF) was the one with the fewest number of data points, specifically 114 points. This value can be observed in Fig. 5.5.

After re-training the models with this new training dataset and generating the new sizing from the test dataset, the simulator was used to produce the real performances. These were then compared with the test performance values, and the results can be observed in Table 5.5.

It is important to note that neither the MLP nor SL produced a single point inside the saturation zone, so they were excluded from this evaluation. Moreover, ResMLP and EoT had a higher percentage of their points outside the saturation, with 60 and 40%, respectively. Nevertheless, both of these models were able to predict the

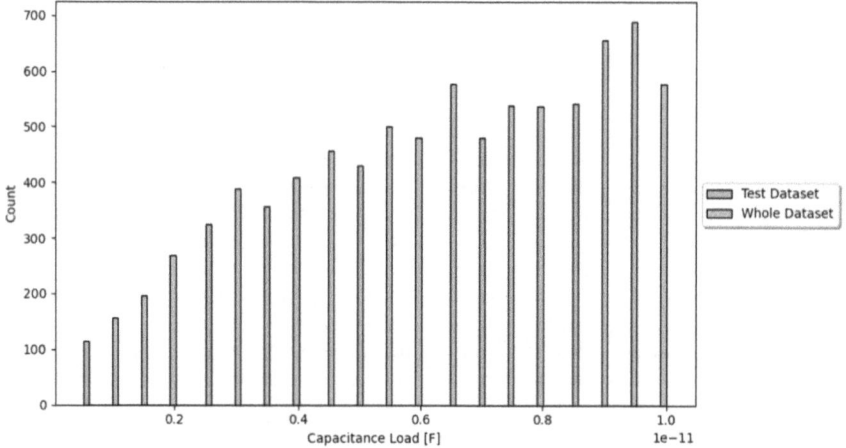

Fig. 5.5 Folded VCOTA C_{Load} value selected for test evaluation

Table 5.5 Simulation test error of the folded VCOTA dataset

Performances	Median error (%)	
	ResMLP	EoT
GBW	10.86	11.06
G_{DC}	3.24	3.90
I_{DD}	4.57	4.55
PM	4.80	4.25
Average	5.87	5.94

context-independent points with an error of around 5%, demonstrating their generative capabilities once again. Surprisingly, both the parameters the I_{DD} and PM exhibited lower errors compared to the random test split for both models.

5.3   Conclusion

In this chapter, the experimental results obtained from the optimized models were discussed. The models were evaluated in five different scenarios: two for the VCOTA and three for the folded VCOTA. These scenarios were chosen to represent a range of different metrics, focusing on accuracy in test evaluation and the generalization of the models through inference tests, which allowed us to assess the models' capabilities and limitations. By comparing these results to a state-of-the-art model, limitations and shortcomings in the approach outlined in the relevant paper were identified. Additionally, by using another SL model, a better baseline for comparison was established through a more traditional method.

The evaluation demonstrated that this approach outperformed these benchmark models in terms of both accuracy and generalization. An interesting challenge was observed, which was shared by both diffusion models and the more traditional ones: the percentage of points falling outside of saturation, particularly for the second dataset.

References

1. Lourenco, N., et al.: On the exploration of promising analog IC designs via artificial neural networks. In: 2018 15th International Conference on Synthesis, Modeling, Analysis and Simulation Methods and Applications to Circuit Design (SMACD). IEEE (2018). https://doi.org/10.1109/smacd.2018.8434896
2. Povoa, R., et al.: A folded voltage-combiners biased amplifier for low voltage and high energy-efficiency applications. IEEE Trans. Circuits Syst. II: Express Briefs **67**(2), 230–234 (2020). ISSN: 1558-3791. https://doi.org/10.1109/tcsii.2019.2913083

Chapter 6
Conclusion and Future Works

6.1 Conclusion

In this work, the area of analog Integrated Circuit (IC) sizing [1, 2, 2, 3], its ill-posed nature, and a novel approach to addressing this problem were studied and implemented. The inverse problem was first analyzed, and previous approaches used to solve it were reviewed [4–10]. After exploring this area, attention was turned to studying a type of generative Artificial Intelligence, specifically the Denoising Diffusion Probabilistic Models (DDPM) model [11, 12]. By studying these seemingly unrelated fields, a way to unify them was found, leading to the development of an approach that uses diffusion model to address this challenge.

6.2 Abstract

In this book, three different models were presented, all of which were implemented and optimized to perform circuit sizing based on specific performance values. The optimization was not carried out using traditional techniques but instead by considering the target performance.

The experimental results indicated that the models successfully sized the two tested circuits with an average median error of around 6%. For the smaller circuit in terms of the number of sizing design variables, all models outperformed the state-of-the-art approach, whose error was over 60% higher than that of diffusion models. In the case of the larger circuit, the supervised approach led to an average error 70% larger than that of most accurate diffusion model. Furthermore, by leveraging the generative capabilities of the diffusion models, points for targets within the dataset were generated, with most of them showing an error below 3%. For more challenging targets, solutions with errors below 10% were found, while the supervised approaches struggled to achieve errors under 20%. When analyzing performance individually, even when the models were not able to sample one point where all the performances were close enough to the performance values of a target, they were able to identify solutions with individual performance values with less than 1% error in regard to each target performance, even for the most challenging cases.

In summary, superior accuracy and generalizability were offered by the diffusion models compared to the benchmark models, and by replicating the state-of-the-art, problems and limitations in that approach were uncovered.

6.3   Future Work

In this section, three different areas in which this methodology could be expanded will be discussed. The first topic addresses an issue encountered in this work, primarily focusing on the dataset. The second topic pertains to the denoising Artificial Neural Network (ANN). Finally, this section—and the book—concludes with a discussion of potential modifications to the diffusion model itself that could be beneficial.

6.3.1   Dataset

The first proposed direction that could be interesting to explore in future work is directly related to one shortcoming observed in these models, the number of generated points outside the saturation. This topic is the most straightforward of the three and relates to the dataset that is fed to the model.

As previously mentioned, the datasets only contain information about circuits within the saturation region, which could limit the model is ability to generalize circuit behavior. This limitation might be causing the high percentage of points outside the saturation region. Since this model, using this dataset, lacks knowledge of the behavior outside this zone, it assumes that any generated point follows the same behavior as the circuit inside the saturation region. A simple way to address this issue would be to add a small set of circuit sizing outside the saturation region to the dataset and use a (or multiple for each transistor) binary variable. This new variable could help guide the model to remain within the saturation zone.

6.3.2   Denoising Neural Network

In this work, three different ANN architectures have already been explored, each building on the previous implementation. As previously discussed, increasing the complexity of the networks appears to improve the stability of the diffusion model. Building on the already developed and tested network, it is believed that replacing the Multi-Layer Perceptron (MLP) with a Convolution Neural Network (CNN) could lead to improved predictions. Considering the translation-invariant nature of the noise added at each time step, a CNN could further refine the model's predictive capabilities. The translation-variant nature of the MLP might explain why good results were not achieved with larger time steps.

6.3.3 Diffusion Model

The final topic discussed in this book is related to improvements to the diffusion model. Two topics that are addressed in this section are: The first one was previously mention and is the inclusion of the re-parameterized of the reversed variance proposed on the paper [12]. This approach introduces an additional parameter for the model to predict and incorporates a new term into the loss function. By using a learned re-parameterize of the variance, the model could be better equipped to learn the distribution more effectively, which can lead to enhanced performance and stability of the diffusion process.

The second topic is related to a common approach in different Machine Learning (ML) models regarding how the different weights of the network are updated: the Exponential Moving Average. This approach creates a copy of the diffusion model, and instead of continuously updating the weights in the network, it updates based on the average of weight of the copy, giving more weight to old information. This approach could help to smooth the training of diffusion models.

References

1. Passos, F., et al.: Enhanced systematic design of a voltage controlled oscillator using a two-step optimization methodology. Integration **63**, 351–361 (2018). ISSN: 0167-9260. https://doi.org/10.1016/j.vlsi.2018.02.005
2. Rocha, F.A.E. et al.: Electronic Design Automation of Analog ICS Combining Gradient Models with Multi-objective Evolutionary Algorithms. Springer, Berlin (2014)
3. Mendes, L., et al.: In-depth design space exploration of 26.5-to-29.5- GHz 65-nm CMOS low-noise amplifiers for low-footprint-and-power 5G communications using one-and- two -step design optimization. IEEE Access **9**, 70353–70368 (2021). https://doi.org/10.1109/ACCESS.2021.3078240
4. Lourenco, N., et al.: On the exploration of promising analog IC designs via artificial neural networks. In: 2018 15th International Conference on Synthesis, Modeling, Analysis and Simulation Methods and Applications to Circuit Design (SMACD). IEEE (2018). https://doi.org/10.1109/smacd.2018.8434896
5. Lourenco, N., et al.: Using polynomial regression and artificial neural networks for reusable analog IC sizing. In: 2019 16th International Conference on Synthesis, Modeling, Analysis and Simulation Methods and Applications to Circuit Design (SMACD). IEE (2019). https://doi.org/10.1109/smacd.2019.8795282
6. Beaulieu, P.-O., et al.: Analog RF circuit sizing by a cascade of shallow neural networks. IEEE Trans. Comput. Aided Des. Integr. Circuits Syst. **42**(12), 4391–4401 (2023). ISSN: 1937-4151. https://doi.org/10.1109/tcad.2023.3282570

7. Mina, R., Sakr, G.E., Nassif, H.: Enhancing transistor sizing in analog IC design using a circuit-focused semi-supervised learning. In: 2023 IEEE 4th International Multidisciplinary Conference on Engineering Technology (IMCET). IEEE (2023). https://doi.org/10.1109/imcet59736.2023.10368264

8. Fayazi, M., et al.: AnGeL: Fully-automated analog circuit generator using a neural network assisted semi-supervised learning approach. IEEE Trans. Circuits Syst. I: Regul. Pap. **70**(11), 4516–4529 (2023). ISSN: 1558-0806. https://doi.org/10.1109/TCSI.2023.3295737

9. Wu, Z., Savidis, I.: Transfer learning for reuse of analog circuit sizing models across technology nodes. In: 2022 IEEE International Symposium on Circuits and Systems (ISCAS). IEEE (2022). https://doi.org/10.1109/ISCAS48785.2022.9937457

10. Leibl, M., Graeb, H.: Optimizer-free sizing of opamps leveraging structural and functional properties. In: 2024 20th International Conference on Synthesis, Modeling, Analysis and Simulation Methods and Applications to Circuit Design (SMACD). IEEE (2024)

11. Ho, J., Jain, A., Abbeel, P.: Denoising diffusion probabilistic models (2020). arXiv:2006.11239

12. Nichol, A., Dhariwal, P.: Improved denoising diffusion probabilistic models (2021). arXiv:2102.09672